RADIO CONTROLLED GLIDING

RADIO CONTROLLED GLIDING

A survival guide for beginners and
intermediate level fliers

DAVE JONES
with slope-soaring material by Keith Thomas
and line illustrations by Roy Garner

ARGUS BOOKS

Argus Books
Argus House
Boundary Way
Hemel Hempstead
Herts HP2 7ST
England

ISBN 0 85242 884 7

Phototypesetting by TNR Productions, 19 Westbourne Rd,
London N7 8AN, England
Printed and bound by LR Printing Services Limited, Edward Way,
Burgess Hill, West Sussex, RH15 9UA

CONTENTS

INTRODUCTION 7

1 GETTING STARTED 9
First steps and equipment choice

2 AVOIDING CONSTRUCTION ERRORS 21
Problem areas and their solutions

3 REPAIRS 51
Common breakages and their repair

4 TRIMMING AND ADJUSTING 59
Setting up the model to fly reliably

5 LAUNCHING 67
Methods and techniques

6 FLYING 83
Methods and procedures

7 LANDING 91
Methods and procedures

8 THERMALLING 98
Lift spotting and use, procedures and
problems

9 SLOPE SOARING (by Keith Thomas) 108
Methods and procedures

10 FOUL WEATHER FLYING 133
How to deal with adverse conditions,
flying to survive

11 THERMAL SOARING COMPETITIONS   144
    Introduction, what to expect, tactics

12 SLOPE SOARING COMPETITIONS   161
    (by Keith Thomas)
    Different types and what to expect

APPENDICES

1   The Beaufort wind scale with   170
    recommendations for thermal soarers.

2   The F.A.I. limits and their application to 172
    current classes of models.

3   Current (1986) specifications for   173
    competition thermal soarers.

4   Possible training schedule/checklist for 174
    beginners.

5   Bungee recommendations.   175

6   B.A.R.C.S. Open Class Rules.   176

7   B.A.R.C.S. 100S Class Rules.   183

8   B.A.R.C.S. Thermal Soaring Achievement 187
    Programme.

9   B.A.R.C.S. Slope Soaring Achievement 189
    Programme.

GLOSSARY OF USEFUL TERMS   192

INDEX   204

DEDICATION

This book is dedicated to the memory of Mac Hurdman, a good friend and a good flier. Also to all my friends and acquaintances in the modelling word for the fun and interest that they have brought into my life. Thank you all – D.J.

INTRODUCTION

It is a great source of sadness to me to see a keen beginner become a disillusioned and disgruntled refugee from soaring simply because he or she has made elementary mistakes. It is hoped that this book will dispel the mists of inexperience, jargon, "misfortune" and heartbreaks that surround learning to fly model sailplanes. In many ways this is the book that I wished that I had been able to read when I was in my first few years of soaring.

The aim of this book is to provide the information and advice necessary for the beginner in soaring to be able to become a competent sport flier; in other words it is aimed at pilots whose knowledge and skill range from that of the raw beginner to the sort of expertise expected of a competent sport flier. There are many things that only first-hand experience can teach you; this book will not allow you to rise from your armchair and be immediately successful. However, it will allow you to tackle the learning task with, hopefully, a greater understanding, knowledge and chance of success.

There are many pitfalls involved with radio controlled soaring, but a great many of them are easily avoidable. Many times I have suffered the slings and arrows of outrageous incompetence, seldom has it been a matter of bad fortune. Good fliers are not fortunate (other than in terms of talent), they have learned their limitations and fly to them. If this book helps you to live by your abilities and expand them whenever possible, then it will have achieved its purpose. *Ut vivatis volate!*, or words to that effect.

You will find some statements that might be considered contentious, and if you feel strongly about anything please write to me via the publishers. However, any such statements are based upon hard and frequently expensive experience. It is perhaps inevitable that this book is a personal view of good practice, and where an opinion is stated I hope that it is obvious that it is a personal opinion. I would not claim a monopoly on the truth: there will be many other solutions that are as good as those presented here, and time will throw up even more. However, I do know that the methods presented here have worked well and continue to do so.

In the cause of brevity it has been necessary to be selective about the items presented and you will find that some of the more traditional modelling skills are not included here, though they can be easily found elsewhere. The reason for this is that it was felt that the content of this book should represent current practice at the time of writing. Within a few years it may well need updating, but the trends seem fairly well set and only the advent of dramatically improved materials, competition rules or equipment should put this book out of date.

Modellers are the biggest scroungers of technology and materials – seldom has anything been developed for our benefit alone. Some items mentioned in this book are not available in the modelling shops, some are even waste materials; acquiring such materials is not particularly difficult but may require some creative scrounging or buying. Wherever possible the sources for unusual materials have been given so that you may find them.

Acknowledgement must be made to the contributions of Keith Thomas and Roy Garner. Keith is as much a slope soaring enthusiast as I am a thermal soaring fanatic, so between us we should have covered all the relevant material that you need to know at the initial and intermediate stages of learning. Roy Garner's graphic and design skills should be familiar to you from his cartoon work in the radio control magazines and the models in the Soaron Sailplanes range, particularly the Sunshine series. It has been a pleasure to work with two such talented modellers.

Finally, a word of warning: never ever take soaring too seriously. In many ways it becomes a way of life, a very enjoyable and satisfying one, but nevertheless something that must be treated lightly and with good humour. Frustration and despondency await those who cannot stand back from this sport and see it for what it is, an intrinsically absorbing, elegant and potentially beautiful interest. I wish you good luck and much enjoyment.

October 1986   Dave Jones

1 GETTING STARTED

All of us started out as red-hot keen types with shiny eyes and a lemming-like urge to spend money, the sort of individuals that salesmen dream about. This is a wonderful state to be in if you are properly advised about your initial purchases and moves; many survive this phase unscathed and go on to feel that the hobby has served them well. Others, however, throw caution to the winds and get their fingers burnt by investing too quickly in equipment that is unsuited to them and their stage of learning. So how do you proceed in a way that will avoid disappointment? Well, since you are reading this book then you have taken the right first step and hopefully you will not have bought anything else yet. The second step is to join a local club.

Being in a club that suits you is half the fun of soaring. Joining a club as soon as possible is the best move you can make. The major advantage for the beginner is the amount of free advice available, advice that is normally not affected by commercial

The 72in. *Gentle Lady* is very typical of the size and style of trainer model recommended.

considerations. The club members will know only too well the models that are suited to the local conditions and terrain, knowledge that will have been gained by hard experience. In addition they will have a wealth of experience concerning radios, launching apparatus, building, finishing and so forth. I am not saying that the model shops will give you bad information – the majority of them give excellent advice to beginners – just that it pays to check your enthusiasm until such advice can be confirmed by consultation with experienced fliers. It is in the interest of the sport that you become a competent flier and it is also in the shops' interest that you are active for several years, so a short delay before you make your mind up will not hurt. It is possible to make the following recommendations without too much risk:

THE FIRST MODEL

The first model must be regarded as a sacrifice to the altar of learning. It will very probably end up either looking very battered, or broken, probably the latter. So it does not pay to take a "pride & joy" attitude to the first model, in fact it does not pay to get too attached to your first three or four models, or any model for that matter. The point that needs to be made here is you must never be frightened of losing a model: if it's too precious to risk then you will be frightened when flying it. This will slow down your learning and the crashes when they do happen will be harder to take. If you are frightened about bending models then it would be better for you to make plastic scale models.

So what should you look for in a trainer? Much depends upon whether you will mainly fly slope or thermal, but in either case simplicity is the key. A simple two-function model is required, that is one that uses only the elevator and rudder for control. Do not be tempted to get a slope trainer that has ailerons (buy it as a second model) as rudder for turn control is far more forgiving than ailerons when you are learning to fly.

Repairability is most desirable for reasons already mentioned. The ability to repair a structure depends to a great extent upon how well you know the structure, and for this reason it is better if you have built the model yourself rather than bought a model pre-built. So if you have some building skills use them, but if you haven't don't worry, as you will learn as you go along.

The size of model that you buy is important, since a small model will "bounce" much easier than a big one. Similarly a light model has less inertia and is more likely to survive a crash. For thermal work something in the 72 inch/2 metre range should be suitable. A 100 inch model will give better duration but it is more likely to suffer damage. Do not be tempted to go for a competition 2 Metre model at this point, as their capabilities may well be beyond yours. Slope work can be achieved easily in light winds with the sort of model mentioned above, but a smaller span of say 60 inches/1.5 metres is to be recommended for good lift conditions and tight landing areas. Some of the vintage (pre-'51) gliders would make suitable trainers.

The decision to buy a ready-built model or build your own is dependent upon your previous experience. The ready-built model should at least have a reliable level of performance and will get you flying in the shortest possible time, but the cost may well be higher than for a kit. Building your own model from a kit is to be preferred if you have some experience of building in your past. There is a great deal of pleasure to be had from building but the extra cost is in terms of your time and the family's patience.

THE RADIO

The second big purchase is the radio gear. Unlike the first actual model it is likely to be used for quite a long time, so this choice is more critical. The model shop should be able to suggest a set of gear to suit your pocket, but club membership really pays off here since your clubmates will have plenty to say about the commercial gear available and will have their experiences to tell. Be careful to sift the objective comments from the personal prejudices. Additionally, the club's instructor(s) may possess a Buddy Box lead, which allows two transmitters to be linked together so that in an emergency the instructor can take over control, rather like a dual control car for learners. Transmitters of different manufacture may not be compatible so this may well be a factor to consider when purchasing.

How many functions? The beginner does not require a radio that does everything except brush its teeth before breakfast, simplicity is the key. Strictly speaking you only need a two-function set for learning, but as this can pose problems as you progress onto more complicated models, a better buy is a three- to four-function set. A four-function set offers the greater flexibility, its two dual axis sticks allowing you to progress into ailerons, flaps, airbrakes and so forth without having to buy another radio. Having a further fifth function that is switchable is very useful for airbrake operation but not essential.

Modern transmitters offer a lot of different options, but during the process of learning most of them would not be used. The option that is well worthwhile is the provision of rate switches. These will reduce the amount of control movement down to a preset minimum. The best way to describe this is for you to imagine that you are driving a car that has steering that only requires two turns from lock to lock, the car would respond very quickly. If the steering were now changed to four turns lock to lock the steering would be slower and far less sensitive. The rate switches have the same purpose and effectively desensitise the controls by reducing the amount of control movement possible. This is most useful since it allows you to convert the model from a docile trainer to a highly responsive model at the flick of a couple of switches, thus you can tune the model to your developing responses and experience. In addition if you have set the model up with too much response you can cut back on the throws with the rate switches without having to mess about with the servos and control linkages. This should only be regarded as a stop-gap measure, though, and you should sort things out properly back at the workshop; a servo on rates has less accuracy.

The other feature worth looking for is servo reversing, usually carried out by the use of switches on the back of the transmitter. Say you arrive at the flying field with the elevator operating in the wrong sense (wrong way round). If the installation will not allow

An example of a 5-channel transmitter with servo reversal, rate switches etc. which naturally cost more.

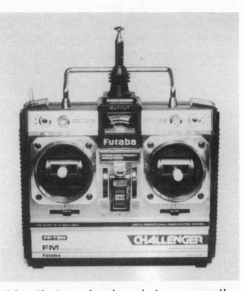

for correction the only thing that can be done is to reverse the servo. Before switching the only answer was to use a "reversed" servo or juggle the wires, but now all one needs to do is to flick a switch. Be warned, though, this is a double-edged sword! It is common practice to use the same radio for two or more models, either by moving the receiver around or by buying an additional receiver for the other model. It can happen that the other model requires a different servo set-up, so that it is necessary for you to reset the switches before flying the other model. All will be well until one day you forget to do the reset, and the result is almost always painful. Right becomes left, and/or up becomes down, and the crash happens so fast that it is usually only when inspecting the wreckage that the cause is found. It's far better if all your models are set up in the same way so that you need not worry about such problems. The situation is easily avoided if you do a pre-flight check before each flight, including a range check at the start of the flying session. Do not be put off if the set does not have servo reversing.

Mixers are not necessary for learning, but they become important with more complicated models that have flaps, coupled ailerons and rudder, V-tails and so forth. If the transmitter offers one it can be useful at a later date, but it is not essential initially. Roll switches and the like are not very useful for any but the most experienced fliers.

Expandability is a feature of some sets. With these you can

add the various functions, mixers etc that you need at a later date as and when they are required. This is quite useful and should save you from having to buy a complicated set when you progress onto more demanding models.

At the time of writing, the computerised radio has just appeared. For someone who started with a basic M series Futaba outfit this type of radio presents all sorts of mind-boggling options and possibilities. For a beginner the sheer complexity of such gear may be too much – more options mean more things to go wrong and simplicity has a great deal to offer. For an intermediate level soarer thinking of entering competition this sort of equipment may well be invaluable.

Batteries require some consideration. The cheapest sets are usually set up for ordinary dry cells, which are satisfactory for earthbound models such as cars and boats but far from worthwhile for any airborne model. The best answer is to use rechargeable batteries, nickel cadmium cells (nicads). There is a price penalty in terms of initial cost but this will soon be recouped by not having to purchase dry cells at regular intervals. The problem with dry cells is that you never know how old they are (age affects the amount of charge in the battery, charge decays with age). Additionally you never know how much charge is left in a battery when it has been used, so you can never tell how much safe airtime you have left. On the other hand with nicads you can, by charging them overnight for some 14 to 18 hours, ensure that they are always fully charged. You will then know roughly how much operating time you have at your disposal. With nicads you are very much in control, with dry cells you never know where you are.

Converting a dry cell outfit to nicads must be done properly. It is *not* good practice to charge up the cells and drop them into the back of the transmitter and into the receiver battery holder pack. What will happen is that after a while the contacts will become worn and sooner or later the transmitter or receiver will mysteriously stop of its own accord. Usually it will equally mysteriously switch itself back on; frequently the jolt of the crash will get the receiver batteries going again and releasing your grip on the transmitter casing may well do the same job. The only way of avoiding these problems is to wire the batteries permanently into packs for the transmitter and receiver and charge them up via proper plugs and sockets. This is a job for the manufacturers' agents, for an unauthorised conversion might void the guarantee on a new set. The model shop may well be able to do the job for you, but if they cannot then someone at your club should know how to achieve the desired result. When buying nicads look for the types with "solder tags" which are much easier to make into packs.

Servos come in sizes ranging from micro and mini, through the standard size to large and giant. Each type has its uses:

Micro-   Used on small models such as hand-launch gliders, tiny planes and for installation in wings. Tend to be expensive and low on power, not suitable for beginners

Mini-   Suitable for most general purposes on gliders

Standard- Designed for general purposes and usually a little cheaper due to volume production, tend to be more powerful than the above

Large-   Designed for heavy work such as retractable undercarriage operation, often modified from standard servos

Giant-   Designed for very heavy duty work.

The best ones to buy are the standard, or the mini type if the budget can run to it. The vast majority of models are designed for standard servos, but the use of minis can save space in a tight installation. Initially you need only buy as many as you need, usually two for the first model. Any further ones can be bought as and when needed and to suit the intended purpose. At the last count my personal collection had reached 24, mostly standards but with a fair range of the rest. Some makes of gear are not critical about whose servo they use so it is possible to fit different plugs to suit your set and thus make use of second-hand gear from another manufacturer. Receivers can be equally as accommodating and it is possible to use a receiver and transmitter from

Various servos commonly encountered include, left to right, linear output, standard, mini, alternative standard, micro and high power.

different manufacturers, but care must be taken and it is good practice to consult the service agents to check compatibility. Crystals and plugs are usually a problem. If you do try this make sure that the whole set-up is thoroughly checked prior to flight, in particular in regard to the range check.

LAUNCHING EQUIPMENT

There are three main types of launching equipment in everyday use: the bungee (Hi-Start), towline, and power winch. The power winch tends to be used for heavy models such as scale sailplanes and fast heavy models such as F3B gliders. The case in other countries is different with winches being common, but in Britain they tend to be the exception rather than the rule. For the beginner then the choice tends to come down to the bungee or the towline, though this requires a runner so one cannot work alone. The bungee is effectively a 500 foot monofilament catapult consisting of a length of elastic material attached to a length of fishing line. The material used for the elastic element determines the effectiveness of the apparatus in different weathers. There are basically two types: "Cloth covered" and "Surgical"; a third, "Solid Rubber", is rarely seen. Undoubtedly the most useful is the surgical type, which uses a special surgical tubing often imported from America and usually somewhat expensive. However, it will give you a long gentle pull and is capable of full height launches in winds of about 8-10mph. It is, however, more prone to abuse in terms of overstretching and may only last two seasons. The "cloth covered" type consists of strips of white rubber bound together with a cotton covering outside. This tends to give a powerful launch but the stretch is released quicker than with the surgical type. Full height launches are possible in higher winds, say 12-15mph. Their big advantage is cost and durability, since the covering tends to prevent overstretching and abrasion.

As a beginner it would be to your advantage to use a clubmate's bungee at first, since you will not need the independence of having your own launching equipment till you can fly solo, and if you do not have it you will not be tempted to try to fly by yourself before you are ready! When you are competent you will need your own bungee. As you go through the mad keen phase, frustration will force you into buying one so that you can take advantage of summer evenings when saner souls are devouring their meals with their loved ones.

The towline is if anything a more useful tool than the bungee; it can give full height launches in lighter weather than the bungee but requires teamwork and practice. Most British competitions use towline exclusively since with set line lengths the resulting launches are roughly of the same height. There are some big advantages: no line tangles (the line is reeled in after the flight), into wind launches (the line is run out for each launch, it is not fixed) and since it is not left lying around for long periods it is unlikely to be accidentally damaged or vandalised by children (it can happen in public parks). A towline is usually purchased by people who are becoming interested in competitive flying.

TOOLS AND ANCILLARY EQUIPMENT

After making your major purchases you will need to buy some specialised tools for modelling, but fortunately most are relatively cheap. The following are essential:

Craft knife – a small one cannot be beaten for fine work, and heavy jobs may be tackled with a heavy duty blade in a Stanley knife or similar. Buy some spare blades when buying the knife – straight slightly angled, straight chisel, and scimitar style blades are generally the most useful.

Straight edges – two are required, a short one of around 12 inches/300mm and a long one of around 3 foot/1 metre. The short one is for small work, the larger for preparing wing sheets and long edges.

Miniature plane – sometimes called a razor plane since they use thick "razor blades" to provide the cutting edge. These little hand planes are invaluable when shaping up fuselages and wings and are very useful for freeing up awkward drawers and such like around the house.

Sanding blocks – two types are required, a cork block for sanding double curvatures such as are found on fuselages and a hard faced block for very accurate single curvature work such as wing leading edges. I personally use a piece of T section aluminium salvaged from a scrap yard for my hard sanding block.

Abrasive papers – get a fair range and do not waste time using a fine paper on a heavy job. You will need to get wet and dry papers for glass-fibre work. Do not overwork a piece of paper, it's a false economy in terms of time and finish.

Building table – can be an awkward item to find. I personally use a heavy desk top made of plywood some 1½ inches thick. Some people have used doors, although modern ones tend to be a bit lightweight. Ingenuity is called for! It must however be flat and be capable of taking pins.

Soldering iron – get a good medium size iron with inter-changeable tips, which will be needed for control installation and electrical work.

Other items of kit can be added as and when required. A small bandsaw, for instance, makes life very much easier, and a sanding disc also saves a lot of time and increases accuracy. Locksmiths' warding files are very useful for hinge slots and slot work. It is best to buy things as and when they are required, since that way you do not collect a load of junk.

No self-respecting flier is complete without some suitable clothing. You need not deliberately buy flying clothes but possession of the following makes life more comfortable:

Winter coat – this should be windproof, pretty waterproof and preferably be thigh length; light weight is an advantage when scaling a slope.

Hat – a great deal of heat is lost through the head and a suitable hat is necessary for slope flying in the winter. A hood will do, or a woollen ski/bobble hat is pretty good and easily jammed in your pocket.

Gloves – essential on cold windy days, when frozen hands can make life miserable. I personally use some skiers' nylon inner gloves with any old leather/leatherette gloves over the top. This combination is very warm and the inner gloves will provide some protection when you are actually flying and need to take the outer gloves off to be able to feel the sticks.

Overtrousers – these are a useful addition to your foul weather kit, for by stopping the wind chill from stripping away your body warmth they make a significant contribution to your comfort. Skiers' tights only seem to work if you are moving and I have no experience of Long Johns! Overtrousers have the big advantage of being less embarrassing to put on than skiers' tights etc. and are usually showerproof.

Sunglasses – absolutely essential, glare and eye strain headaches are no fun at all. Personally I do not like polarising lenses although they are perfectly satisfactory. Straight-forward light-filtering lenses are, I feel, better, since they do not have such a dramatic effect upon sky tones. Many will probably disagree, and it's a matter of preference. Most sunglasses are usually not dark enough for comfort in bright sunlight, so for flying go for the darkest pair you can find and reserve them for the purpose if at all possible.

Slope flying in a good blow can be awkward due to the wind swirling around the glasses and the answer to this is a pair of skiers' goggles, either already tinted or worn over your glasses.

Boots–waterproof but preferably lightweight. Walkers' boots are ideal, Wellingtons will suffice. Remember that you may well need to run with a towline or to maintain sight of your model.

Stopwatch – particularly useful for thermal soaring. A proper stopwatch, preferably digital (it saves arguments) is the best choice. A wristwatch with a stopwatch function can be used but they are awkward and I did see a model crash once whilst the flier was struggling with his wristwatch.

All the clothing tends to be accumulated over the years as experience is gained of how miserable it is to be on a field or hillside when it is freezing cold, windy and possibly raining. You might think that you will not be so foolish as to go out in such conditions but enthusiasm is a great spur and once hooked it will be hard to stay in the house on a flyable day. Even I get caught by the bug from time to time, despite having carefully nurtured a reputation for being a "fair weather flier"

INSURANCE

Third party insurance is essential, and your club may have its own arrangements or be associated to the SMAE and its related insurance policy. If not, join a well-known insurance scheme such as the MAP policy as soon as possible, and put the above items on your household policy list.

So that just about sums up the short and long term expenditure. You are still interested? Let's look at how you avoid making a mess of the model.

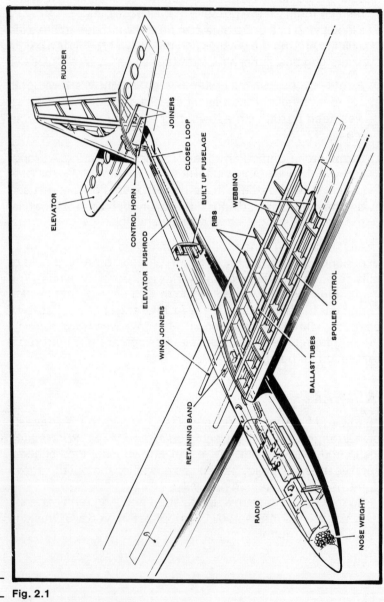

RUDDER

JOINERS

CLOSED LOOP

BUILT UP FUSELAGE

ELEVATOR

CONTROL HORN

ELEVATOR PUSHROD

RIBS

WEBBING

WING JOINERS

SPOILER CONTROL

BALLAST TUBES

RETAINING BAND

RADIO

NOSE WEIGHT

Fig. 2.1

2 AVOIDING CONSTRUCTION ERRORS

Having got through the first hazards, the next big stumbling block is the production of a flyable model. This breaks down into two major areas: Airframe Construction and Control Installation. Providing that these areas have been competently handled the learning process should proceed with few delays, but if they haven't then no end of trouble is in store. These notes must of necessity be brief and generalised, since model building is a vast subject and space limitations prevent me from going into too much detail. Each model will have its own building methods and the following are points that require additional care and have been known to cause problems.

The structure that we call an airframe consists of several major groups of components, each of which has its pitfalls; most mistakes are boringly predictable. Models are so diverse that it is difficult to be precise about recommendations of good practice. The best procedure is to have a competent modeller check your work as it progresses, and preferably at stages where corrections may be made. It is no good checking over the rudder and elevator linkage, for instance, when the fuselage has been completed and major surgery may be required for corrections to be made. Think ahead.

FUSELAGES

These fall into four main groups: vacuum-formed, plastic moulded, wooden and glass-fibre (GRP). Vacuum-formed fuselages are used on some ready-built models and generally they are pretty successful, although they can be a little heavy. There are two minor problems with them, repair and cold weather. Low temperatures can make the thermoplastic material brittle, rendering it liable to shattering if landed hard on a sub-zero day. Generally the only major work needed is to fit the radio gear; more on that later.

Injection-moulded plastic fuselages are used by several manufacturers, usually German. These are incredibly tough and usually a bit expensive, though their toughness soon pays dividends in that they do not break very easily. I once put a semi-scale 11 footer into soft ground from 500ft, vertically, and all it did was distort the nose and mess up the canopy and pushrods. The thoroughness (and expense) of these kits is worthwhile for those looking at a second model. Follow the instructions and this type of fuselage poses few problems.

Wooden fuselages are usually of the "doublered box" type, probably the simplest way of producing a suitable structure. Here, though, there are lots of possible pitfalls and ways of improving the end result. With any kit, follow the instructions carefully, the designer knew what he was doing (even if he does not express it very well), but add the following to your building notes. Plywood is usually used for the doublers and it can some-times have release agent on it from the time that it was manu-factured. It pays to sand both sides of any ply component lightly so as to get a better key for the adhesives and remove any traces of release agent. It is not a frequent problem but one that can lead to the fuselage peeling apart if not dealt with at the building stage. When fitting doublers to fuselage sides the use of any-thing other than an epoxy or wood glue is to be avoided. Contact glues can come unstuck with age and repeated impact. Wood glues, although they are slower working, have the distinct ad-vantage of giving a lighter joint since they cure by the evaporation of their solvent. Epoxy may very well give a stronger joint but it will be heavier (no solvent to evaporate) and the end result may well be a joint that is much stronger than the surrounding material, not a disadvantage but not necessary.

The nose area of a wooden fuselage takes a tremendous battering and a common result of a sudden arrival is that it splits around the front end. A useful precaution is to put a layer of glass cloth in the equipment bay from the nose to the former at the leading edge of the wing, 4oz. cloth is adequate. This substanti-ally reinforces the area and although it adds weight it does so in an area that usually requires additional weight to achieve flying trim. Do this before installing the servo fittings.

A frequent problem is the production of an inadvertently bent (banana) fuselage, and particular care must be taken to avoid this. The use of a jig is to be recommended, as is the prebending of the fuselage sides (steam and pin in the desired shape to dry). A useful dodge is to mark the centre lines on all the formers and

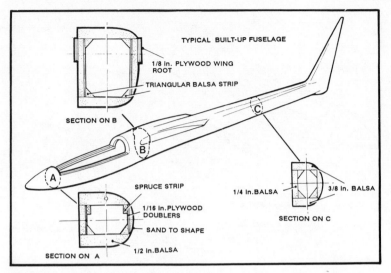

TYPICAL BUILT-UP FUSELAGE

1/8 in. PLYWOOD WING ROOT

TRIANGULAR BALSA STRIP

SECTION ON B

SPRUCE STRIP

1/16 in. PLYWOOD DOUBLERS

SAND TO SHAPE

1/2 in. BALSA

SECTION ON A

1/4 in. BALSA

3/8 in. BALSA

SECTION ON C

Fig. 2.2

use these to line the fuselage up by sighting along them. If despite all your precautions the fuselage is bent, break the formers out and try again. A bent fuselage will try to turn one way all the time, requiring opposite rudder for straight line flight (inefficient), and will turn well one way but not the other.

A common mistake is to strengthen every item on the model and end up with a grossly overweight machine that would probably be highly appropriately described as a "brick". Try for a light, strong structure and only reinforce those areas where breakages are known to be likely to occur. Figures 2.1 and 2.2 illustrate a fairly typical built-up wooden airframe.

Glass-fibre fuselages come in two common types, complete mouldings or pod and boom mouldings. Both will have a canopy opening of some sort and this can be the first problem area. In a crash the canopy opening is very likely to be the area that breaks up on impact. This is due to the fact that it introduces a weak area into the structure. It is not a bad idea to put another strip of heavy glass cloth around the edge of the opening before fitting out the equipment bay. For an adhesive use the same resin that the fuselage was moulded in, polyester or epoxy, for all these strengthenings.

Another useful way of increasing the strength of the cockpit area is to instal a "servo tray" permanently in the fuselage, glassing it in place with a strip of 4-8oz. chopped strand matting (figure 2.3). This can, however, introduce stress points immedi-

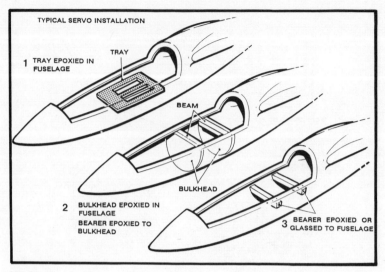

TYPICAL SERVO INSTALLATION

TRAY

1 TRAY EPOXIED IN FUSELAGE

BEAM

BULKHEAD

2 BULKHEAD EPOXIED IN FUSELAGE
BEARER EPOXIED TO BULKHEAD

3 BEARER EPOXIED OR GLASSED TO FUSELAGE

Fig. 2.3

ately in front of or behind the tray and such areas may require additional strengthening. Another good idea for servo mounting is to glass-cloth bulkheads into the cockpit area fore and aft of each row of servos or individual servo (figure 2.3). These installations tie both sides of the fuselage together and prevent it from distorting upon impact. They will not save the fuselage in a very bad crash but will prevent it from suffering major damage in a minor to medium strength incident.

The fuselage will tend to be crushed when the model lands suddenly and the wings swing forward, which can cause splits around the front of the wing root area. The easiest way to eliminate this problem is to fit a crush beam, made from ⅜in. square balsa or ¼in. round hardwood dowel, between the wing roots at the leading edge of the wing.

With a complete moulding you are at the mercy of the manufacturing quality of the moulding. Most are very good but a broken tail end can involve some tricky surgery. For this reason, and their inherent versatility, I personally prefer pod and boom fuselages. These consist of a fuselage moulding, a boom tube and a fin moulding. Because the boom length can be altered it is possible to build fuselages of various lengths to suit the wingspan that we require. For instance, the *Proton* pod has been used on models ranging from 2m up to 160ins. In addition to the canopy areas already mentioned there are three points that benefit from a little extra attention. The boom tube can split along its length and each end should have a yard or so of glass string "whipped" (wound) onto each end and stuck with slow setting epoxy; for glass string pick apart a piece of 4oz. glass-cloth. The spigots that the boom fits onto on the pod and fin mouldings provide another weak point, but an extra layer of 4 to 8oz. glass-cloth about 1 to 2ins. long on the inside of the spigot area of the moulding will usually cure this problem and save repairs later.

There is something of a debate concerning the relative merits of the different resins and glass-cloths to use for moulding. Epoxy is definitely less brittle and thus somewhat more durable. Theoretically, woven glass-cloth is superior to chopped strand matting, but I have found little difference in everyday use, and anyway any GRP system will fail if the impact is strong enough.

Your choice of fuselage type will depend upon your preferences. Generally wooden fuselages are to be preferred on light models and small gliders as they can be built lighter and strong enough. Large, and heavier or faster, models benefit from the use of a glass-fibre fuselage. The differences in terms of building time are not very great, in terms of durability I personally feel that glass-fibre, if well engineered, has advantages.

SERVO INSTALLATION

Servo installation is a common problem area, and the mistakes are usually in terms of gross over- or under-engineering. The servo needs to be firmly held in place but it does not need to be pinned down like a captured psychopath, nor allowed to roam around. A simple pair of ¼in. (6mm) square servo rails are usually adequate if a tray is not being used, secured firmly in the fuselage (figure 2.3). They can be glued onto the bulkheads if these are fitted. To locate the servo simply screw it down with a small woodscrew and a small washer, tightened up until the rubber grommet starts to swell, which should be adequate for all normal purposes. There is no need for a display of ingenuity, save it for later. Figure 2.4 shows a typical installation.

RADIO GEAR PROTECTION

The rest of the radio gear needs to be protected from shock and a common mistake is to pack it in with soft foam, which may look right but is of little value in a crash. Any padding needs to be rigid, and if you can squash it more than 50% by squeezing it between your fingers then it is of no use. Similarly, if it does not recover its

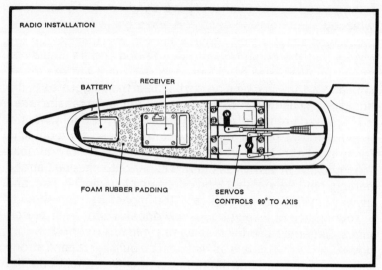

RADIO INSTALLATION

BATTERY

RECEIVER

FOAM RUBBER PADDING

SERVOS
CONTROLS 90° TO AXIS

Fig. 2.4

thickness after squeezing it is not to be recommended. For this last reason polystyrene foam is not very good, since it may protect the radio gear but it may allow it to move to a more forward position and stay there, thus affecting the model's trim. The best crash padding that I have come across is a white plastic foam which looks like polystyrene but has a greasy feel to it; this

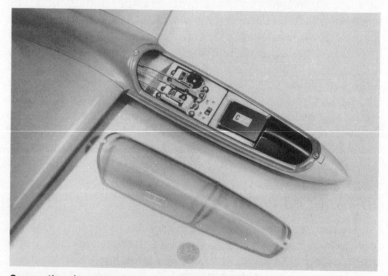

Conventional arrangement of equipment in a *Solitaire* model. Batteries are ahead of the receiver beneath the black area.

cuts and shapes beautifully and is ideal for our purpose. It is often used in the packing of heavier and more expensive electrical goods and most that I have used was ex-computer packing. A good substitute is heavy duty reconstituted upholstery foam.

The installation process is simple. Cut a piece of foam to the length of the equipment bay and trim it down along the bottom and sides until it is a snug fit. Put the block in the bay and mark off the top, trim to size. Now offer the receiver up to the block, mark round it, and excavate a hole the depth of the receiver. Instal the receiver and cut any slots needed for the servo leads, etc. If the equipment bay is large the switch can be sunk into the foam in the same manner. In very tight fuselages the alternative is to use separate pieces of foam.

The receiver battery presents a different problem, as it has the greatest inertia and thus needs the stiffest foam possible. The usual place for it to be installed is the nose of the model, a part of the fuselage that is usually tapering to a point. Two problems arise: packing and positioning. The packing usually has to be made up from strips since the space available frequently will not allow a block of foam to be fitted. The positioning can be critical, as the battery must be as far forward as possible to minimise the trim ballast but it must not be allowed to wedge itself up in the nose of the model. If it can wedge itself it can easily split open the nose or distort it if it is a glass-fibre fuselage. The usual method is to put the nose ballast in first, either as sheet lead or bits of lead in an adhesive, then try the battery for fit. If it does not wedge all is well, but if it does then one can either put a balsa spacer in and then pack the battery or use very dense foam thick enough when squashed to prevent wedging. An alternative would be to make the battery pack up as a "pyramid" with three batteries at the bottom and one tacked on the top.

A minor point to consider is how to remove the radio gear. Some installations allow easy access but removing a battery from the nose of a model when all that is visible is the connecting lead can be a problem. Pulling on the lead is to be discouraged, so the answer is to put a strip of electrical tape around the battery and make it into a tag long enough for you to get a good grip, which does make life very much easier.

CONTROL RUNS

The worst mistakes are made when installing the control runs. Getting these right is absolutely crucial. I have seen too many evil-flying models with awful control runs and conversely I have also seen the same models transformed with a little work. The enemy of any control system is slop, backlash or excessive end float, call it what you will. The rudder is not too critical but it is essential that the elevator linkage be as slop-free as possible, even more so with an all-moving tailplane. There are three basic systems used, Cables (Bowden, Golden Rod, etc), Pushrods and Closed Loop, and all have their little quirks.

Cables are the easiest to get wrong and also the most commonly used in the sort of kits that a beginner is likely to buy. The first mistake is to allow too much of the cable to extend beyond the end of the cable outer. (Figure 2.5 is worth studying at this point.) This lets the exposed cable bend when it is under compression (being pushed) and can severely limit the amount of control surface movement produced. In addition any aero-dynamic load can then deflect the surface from the desired position. You may get away with it on the rudder but if it happens to the elevator it will almost certainly lead to a crash eventually, frequently when you are travelling fast. The best procedure is to leave the cable outer overlength and then cut it back when the controls are being set up. When the cable is pushed there should just be enough cable outer cut back to allow for the full travel of the servo to operate, including the trim movement as well. When braided wire cable is used it is a good idea to remove the cable and run solder into the areas that are exposed beyond the ends of the outers, which will stiffen up the cable and help prevent flexing. If the end still permits bending when the cable is com-pressed, try securing the outer at the very end with a balsa block.

The other big problem with cables is tethering the outer tube in the fuselage. If it is not tethered at regular intervals along its length, or better still along its whole length, it will have the same effect as mentioned above: compression loads will make it bend and cause lost movement and slop. With wooden fuselages it is possible to build the cable outers into the structure, either in the fuselage sides or through the formers and special sub formers. With glass-fibre fuselages the problem arises that you cannot easily work down the fuselage, so you must stick it to the side of the boom or mount it through some sort of bulkhead. Figure 2.5 may help. Glueing a plastic outer cable is not so easy, and though it can be abraded to roughen it the best dodge is to put self adhesive label material over the area to be glued and then use the paper surface as a bonding area. Cable linkages are reason-ably good if carefully fitted but it is a struggle to eliminate the slop. I have seen some that were excellent but many that were indifferent. The cable linkage has one big advantage in that it can negotiate bends and is thus inherently simple.

Pushrods have much to recommend them, as they are simple and, if constructed properly, not liable to bending and lost movement when compressed. They do pose a couple of restrictions though; firstly they require a straight uninterrupted run from the servo to the surface linkage and may thus not be feasible in some fuselages, and secondly they cannot go around corners without the use of a bellcrank system. Hence the popularity of cables in kits. There are three basic problems to overcome: clearances, connectors and flexing.

A pushrod should (just as in the case of the cable) have as short a linkage between the end of the rod and the servo arm or control surface connection as possible. The limiting factor here is stopping the rod itself from fouling on the top of the servo when it is pulled over the top of it at full trim. Figure 2.5 should explain this. The connectors should allow as little slop as possible, which means that the holes in the servo arm and control surface connections should be a tight but not stiff fit for the clevises; they will free up with wear. This applies to all linkages. Do not be tempted to use ball joints for pushrods or cables, for though they are slop-free they will allow the servo arms and surface linkages to bend, and thus allow some lost movement when the surface is aerodynamically loaded.

Fig. 2.5

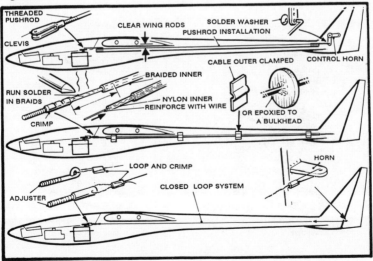

Pushrod flexing has the same effect as an untethered cable outer, but it can be overcome with careful choice of the pushrod material. This will depend greatly upon the length of the pushrod and the size of surface that it is required to control, but generally the best idea is to take the selected material and hold it at the desired length and compress it along its length. If it bends easily then a thicker section is required, but if it is very rigid see if a lighter section can be used. If at all possible instal the connecting rods in the end of the pushrods so that they are in line with the centre line of the pushrod (rather than being attached to its side), this reduces the chance of flexing occurring in flight. Figure 2.6 shows several alternative pushrod ends. The material used for the pushrod should vary with the fuselage construction, balsa for a wooden fuselage, glass-fibre for a glass-fibre fuselage (use the tip blanks from fishing rods, available from fishing shops that repair rods, ask for offcuts). This will allow the rod to expand and contract at the same rate as the fuselage and stop the model going out of trim as the temperature changes. If a very long pushrod is anticipated it is advisable to use some form of guide at the middle of the rod to prevent undue movement. Test for strength by the same test but hold at the end and centre point. Always aim for lightness and the maximum strength to weight ratio.

Closed loop systems are most suited to operating the rudder. Many experts will use such systems on the elevator but the

Fig. 2.6

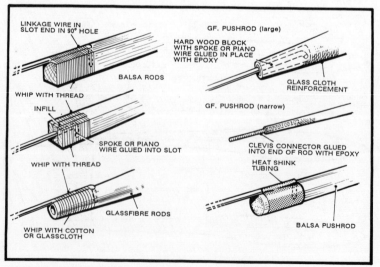

LINKAGE WIRE IN SLOT END IN 90° HOLE

GF. PUSHROD (large)

HARD WOOD BLOCK WITH SPOKE OR PIANO WIRE GLUED IN PLACE WITH EPOXY

BALSA RODS

GLASS CLOTH REINFORCEMENT

WHIP WITH THREAD

INFILL

GF. PUSHROD (narrow)

SPOKE OR PIANO WIRE GLUED INTO SLOT

CLEVIS CONNECTOR GLUED INTO END OF ROD WITH EPOXY

WHIP WITH THREAD

HEAT SHINK TUBING

GLASSFIBRE RODS

WHIP WITH COTTON OR GLASSCLOTH

BALSA PUSHROD

degree of care and engineering needed are such that I personally would not recommend them to beginners and average pilots. Closed loop systems are delightfully simple and effective provided that their main limitation, their geometry, and the fixings are correctly dealt with.

The closed loop system works by running a thin wire of some sort down each side of the fuselage and to each side of the control surface. Figure 2.7 shows the general layout. The wires are attached to each side of the servo output and when the servo is moved it will pull on one wire, putting it under tension, and loosen the other wire. Only the wire that is under tension is doing any work, the other is idling, for which reason the closed loop can be called a pull-pull system, whereas both cables and pushrods are push-pull systems. You could compare it to the reins of a horse, pull one and it will pull the horse's head to one side, loosen that one and pull the other way and the horse's head will move the other way.

The main limitation of a closed loop is that the centre position of the control surface and the degree of freedom from slop depend upon the tension in the control lines. The problem here is that the lines are connected to the servo and any tension is felt by the output bearing of the servo as a side load. While most servos will take the load up to a certain point ("ball bearing" servos are much better at this), they will wear out quicker, particularly around the output shaft area. So the tension needs to be kept to a

Fig. 2.7

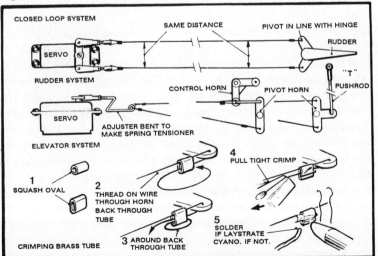

Tail unit of a *Solitaire*. **Ruddér uses closed loop linkage (servo end visible in photo on page 26) and all-moving tail normal push-pull.**

minimum, but even a little tension will take out all the slop in the linkage. However, another factor then comes into play – the lines will have a little stretch in them and it will be possible to deflect the surface by pushing it and thus stretching the lines. Aerodynamic forces will do the same thing, so that some movement will be lost when the surface is loaded. For the rudder this is not a problem and may even be an advantage at high speed (less twitchy), additionally, the rudder is always being knocked about on landing and the springiness of the surface may well prevent damage to the surface and the servo. For elevator control a connection direct to the servo will have the same drawbacks and no flying problems may be encountered until the model is flown fast during the descent from a strong thermal or more probably when penetrating on a windy day. The model may, due to the springiness in the control lines, run out of elevator power. The result may be an ever-steepening dive commonly referred to as "tuck under", although there are other causes for this pheno-menon. My own experiences were painful and I would not like them to happen to others. Because of all these experiences and factors the majority of competition fliers use closed loop for the rudder and pushrods for elevator control; ailerons pose different problems, usually overcome with pushrods or cables.

The geometry of the linkage requires care and figure 2.7 shows the recommended practice. A little inaccuracy will be absorbed by the springiness in the control lines but a large inaccuracy will either slacken the idling line far too much or bind up the system and overload the servo.

The connections are critical and if they are sloppy the linkage will need tightening all the time. The easiest material to use for the control lines is braided control-line model wire (Laystrate), the heavy duty grade. This can be doubled back through the control horn or servo connector, twisted, and simply soldered; any loosening should then be due to the servo settling on its mountings. The other commonly-used material is available from fishing shops as trace wire. This is similar to control-line wire but has a plastic coating and the fixing method involves crimping it with a piece of brass tube. The diameter of the brass tube should be just sufficient to allow three pieces of the wire to pass through when the tube is gently compressed to an oval shape. The brass crimps should be about 3 to 4 mm long and can be made by running the tube under the blade of a craft knife until the tube is deeply scored and snapping off the pieces. It is a good idea to anneal them by stringing them on a piece of wire and passing them through the flames of a gas burner for a while; the crimping will work-harden them. All you need to do now is slip the crimp on the line, pass the line through the control horn, etc. double the line back through the crimp, pull it tight, slip the crimp up as far as possible, and squash the crimp with a pair of pliers. Check by pulling hard on the line to remove any slack, recrimp if necessary. Figure 2.7 should elucidate this process.

All control runs should be as free running as possible, any binding or tight spots in them should be eliminated. Failure to do so will lead to one annoying defect, inconsistent centring, plus extra load on the batteries and potential lock-up of the surface in bad cases.

Inconsistent centring (sometimes called double centring) can be spotted by moving the surface one way and releasing the stick, noting the rest position, and moving the servo the other way and noting the rest position again. If the system is good the surface should always come to rest at the same position, if it does not then the centring is inconsistent. The results of this problem depend upon whether it is the rudder, elevator or ailerons that are affected. With rudder it is not too critical, since what will happen is that the model will not consistently fly in a straight line. A small correction to the right will start it gently turning to the right, the following correction to the left starts a gentle turn to the left and so forth ad nauseam. The same thing happens if the ailerons are affected: the model wants to roll

B

gently one way or the other. After a while the flier will develop a strategy where each movement is followed by a dab of stick in one direction so that the surface is centred on a known position. Now all this nonsense can be lived with to a certain degree, but if the elevator is affected then trouble is in store. Good reliable pitch stability is essential for any model and particularly so for a trainer. Stall control is one of the most crucial things to learn and it will be very difficult to learn it if you are forever fighting to maintain a known centring position and trim. The model will typically either be stalling or shallow diving, and getting it to fly "straight and level" will be very difficult.

The extra load that stiffness in the control lines causes will drain the batteries much more quickly and thus shorten the available safe flight time from each full charge. In extreme cases the additional load on the servo will be such that an aerodynamic load may stop the servo moving altogether (i.e. stall it), which can be a trifle embarrassing.

TOWHOOKS

The final item on the fuselage that needs consideration is the provision of a suitable towhook. Again there is a choice and it basically depends upon whether you use a bungee for launching most of the time. For bungee launching a releasable towhook has advantages. It is possible with a bungee on a windy day to hang on the line like a kite, when the bungee will stretch until it finds a point where it balances the pull. This point can be quite a lot higher than if the bungee were unstretched. With a releasable towhook it is possible to take advantage of this and let go of the line when it is at maximum stretch and height.

Installation depends upon which type is being used. The usual one requires a slot for the mechanism in the bottom of the fuselage which is then reinforced with a piece of 1/16th ply (wooden fuselages) or an extra layer of glass-cloth and a couple of thin blocks for the retaining screws (glass-fibre fuselages). Figure 2.8 shows the arrangement. Do not overdo the strengthening – it is rare for the bottom of the fuselage to pull off on tow! The towhook needs a servo to operate it, ideally one dedicated to the job (an old tired servo is adequate), or one that is already doing another job. The safest one to use is the airbrake servo but it must operate the towhook release without pulling the airbrakes up too far, which means that the towhook linkage has to be a bit hair-trigger in its operation and the servo must be

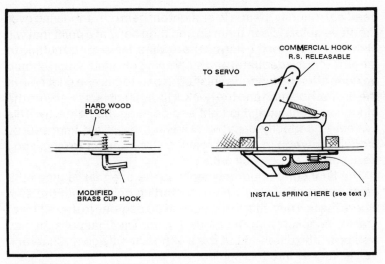

Fig. 2.8

powerful enough to handle the extra load. If airbrakes are not used then the job can be done by connecting the towhook up to the elevator servo so that it operates on DOWN elevator. The release should occur when more than half down elevator is used. This is a safer situation than if up elevator is used since the model is being placed in a position where it is likely to have flying speed and can be brought under control. The only disadvantage is when the line has no tension in it, when if the towhook is dirty it may stick and not release the line. The way round this one is to spring load the hook so that it always opens whether line tension is present or not. A tiny spring (dismantle an old throwaway cigarette lighter) positioned within or around the retention stop at the rear end of the housing does the job well. When setting up the release point always hang a pair of pliers or some such from the hook to simulate line tension.

The type of hook used most is the straightforward fixed tow-hook, which has the advantage of being simple, trustworthy and cheap. Whatever type you use you will need a block of hardwood or ⅛th ply installed in the fuselage to take the screw or screws, and no slot is required. Plastic moulded towhooks are satis-factory for lightweight models but any model that launches fast is better equipped with a metal towhook. There are several designs possible and a little ingenuity should produce an acceptable out-of-the-rut alternative, but in my experience the best one is the simplest, viz the modified cup hook, figure 2.8. Your local hard-ware store should be able to supply "brass" cup hooks. All you need do to modify them is to straighten them out (annealing over the stove helps), clean them up and then bend at a point that will allow clearance for the ring on the towline. Finish off by cutting to length (1 cm is adequate) and smoothing off. Installation should be done after the model is finished. It only requires a pilot hole in the desired location and the hook is then simply screwed into the mounting block until it is tight and pointing the right way. This system is perfectly capable of taking the strain imposed by models that require 100lb breaking strain line for towing, and I have yet to see one pull out.

The final point about towhooks is the one that requires the most care, positioning. The best starting point is with the spot where the tow ring sits on the hook at 30 degrees to the centre of gravity, checked with the model's plan. The final position may well be a little different but it is a generally safe spot. Chapter 4 delves further into this subject.

TAIL SURFACES

These need to be light and strong, the degree of their strength depending upon the speed range of the model. Floaters can have lightweight tails, F3B models need torsionally stiff and strong tail surfaces. However, it must be noted that ⅓ to ¼ of an ounce of weight saved at the tail end will save an ounce of nose ballast, so a fine balance between weight and strength needs to be struck. The majority of models will come out tail heavy. Kit manufacturers and plan designers have normally done their homework, so following the instructions should be sufficient.

In the interests of saving weight it is advisable to use PVA (woodworking) glue on the majority of the joints, and epoxy on highly stressed areas like joiner tubes. Keep the joints accurate and avoid using any excess glue.

The tailplane (horizontal stabiliser) should be thought of as a small wing flying in formation with a larger wing, and all the things that apply to wings apply with equal force. It should be flat, unwarped and consistent on both sides. There are two basic types of tailplane: all-moving and separate elevator. The separate elevator tailplane is theoretically the more efficient of the two and is the traditional choice. Construction is simple and straightforward and from a flying point of view it is more forgiving of control run errors. There are only a couple of problem areas, decalage and control surface stiffness. Decalage is the angle that the tailplane flies at in relation to the angle of attack of the wing.* This is a problem that can only be sorted out at the flying field when trimming the model, see the next chapter. The control surface should be free to flap or vibrate when the surface is disconnected from its control run, as any stiffness in it will have the same effects as a stiff control run.

All-moving tailplanes, as their name implies, are the ones where the whole surface moves to act as an elevator. These have two major advantages; decalage can be dealt with by adjusting the trim on the transmitter, and they can be completely removed from the fin, thus making transport less of a problem. However, they are not so forgiving of building inaccuracies, particularly with regard to control runs. They can be easily blown back by the air or flutter if there is springiness or slop in the system. This is a serious concern with fast models.

Construction is relatively straightforward but there are a few things that can be done to improve matters. The tailplane itself benefits from strengthening around the pivot and actuating rod wire. This can be achieved by putting a "dart" of 1 oz. glass-cloth over the root area on the top and bottom on a solid balsa tailplane (figure 2.9 shows the high stress area) which need only come out to a third span. The extra weight that this adds can be recouped by lightening the tailplane half with a series of holes in the low

*Although this is incorrect – decalage is the difference in incidence angles of the wings of a biplane – it has become a commonplace interpretation with radio fliers.

Fig. 2.9

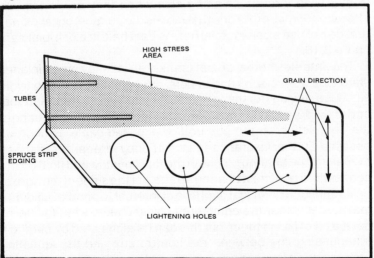

HIGH STRESS AREA

GRAIN DIRECTION

TUBES

SPRUCE STRIP EDGING

LIGHTENING HOLES

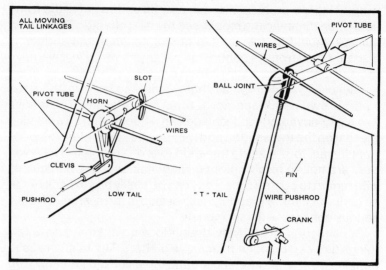

Fig. 2.10

stress areas, as in figure 2.9. Built-up tailplane halves are usually satisfactory in their "as designed" state.

The pivot wire, or more accurately its brass bush in the fin, is likely to become worn and floppy, which is not too serious but looks unsightly. If at all possible permanently fix the pivot wire in the fin; it is less likely to become floppy and it is a much stronger arrangement. It makes transport a little more difficult but a couple of foam blocks jammed on the wires will prevent them from poking through anything else when they are in the car. The actuating wire (the one that moves the surface) on tailplanes that do not use a commercial horn is best held in a ball joint as in figure 2.10.

The internal horns or bellcranks on all-moving tailplanes should be free-moving and special care must be taken to avoid slop. Use clevises for the connections, preferably the metal type, and make sure that the holes are a good fit. Keep the overhang of the connection on the pushrod or exposed cable as short as possible. Further insurance against a floppy wire can be arranged by making the bearing blocks in the fin eliminate end float on the control horn by almost bearing up against the sides of the horn or bellcrank. One point that might be of interest to people building a fast model is that the commercial plastic all-moving tail horns have a bit of flex in them, but this can be eliminated by putting a triangulating link between the control arm and the actuating arm. A piece of wire is adequate, see figure 2.10.

The fin and rudder (vertical stabiliser) should also be thought of as a small wing but this time flying at right angles (90°) to the wing. There are only a few points to consider: weight, ground clearance, aerodynamic balancing and strength. The weight should, as ever, be kept to a minimum and an open structure for non-stressed areas is to be recommended, such as the fin area above the tailplane pivot area and the rudder itself. Light weight in a control surface is worth chasing for another reason – a light control surface is less likely to flutter at high speed, its resonant frequency being high. Ground clearance can be arranged for by sweeping the rudder up at the bottom. Additionally if the model swings round on landing the cut-back will prevent it from being too badly deflected, see figure 2.1. It is also helpful if the bottom hinge is placed as close to the bottom as possible. Strength in a rudder is needed at the bottom, to take landing loads, and at the pick-up point for the control horns.

Putting an aerodynamic balancing horn on a rudder is a mistake for the vast majority of sport models, since it usually leads to rudder flutter. If the kit or plan specifies such an arrangement modify it so that the balance horn becomes part of the fin. You will find no difference in the power of the rudder. The ordinary arrangement has less drag and is less susceptible to landing damage.

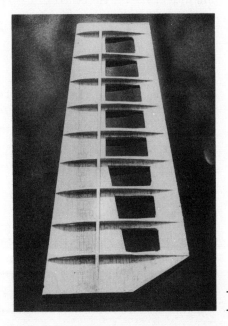

A lightweight tailplane half built on the sheet core system, with lightening holes in areas of low stress.

HINGES

The installation of the hinges is important, get it wrong and the surface will be stiff. There are three basic types to consider: pin hinges, Mylar strips and covering material hinges. Figure 2.11 shows the common types and their installation.

Pin hinges are like small cigar box hinges and are normally made from plastic. These are best suited to control surfaces that require maximum freedom of movement over large deflections, such as rudders, flaps and, to a lesser extent, ailerons. There are quite a few manufacturers and most are perfectly adequate, but try to use the ones with a metal pin in the hinge, as they are stronger and move more freely. Installation is relatively simple: mark out the centre line of the rudder or elevator and the post or spar that it will be installed in. Decide the hinge positions and mark off on both the surface and its location. Cut a slot with a knife and file out the slot to the width and depth of the hinge. Cut a "V" in the surfaces so that the hinge sits flush in the slots with the hinge pin in line with the hinge line. Then chamfer the control surface to allow it to move from side to side.

The next part is tricky to get right. Put glue into the slots on the control surface, push it down and wipe off any excess that may be in the "V". Instal the hinge and check that no glue has got into the moving parts. Allow to cure; epoxy or the rubbery adhesive

Fig. 2.11

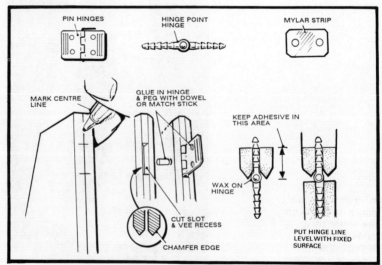

PIN HINGES

HINGE POINT HINGE

MYLAR STRIP

MARK CENTRE LINE

GLUE IN HINGE & PEG WITH DOWEL OR MATCH STICK

KEEP ADHESIVE IN THIS AREA

WAX ON HINGE

CUT SLOT & VEE RECESS

CHAMFER EDGE

PUT HINGE LINE LEVEL WITH FIXED SURFACE

known as "Modellers' Glue" seem most suitable. When dry, offer the surface up to the post and dry fit it to check for free movement and alignment. Then repeat the above process, but be most careful to avoid any glue contamination. If it happens it can be picked out if you are lucky, but often it cannot. Some hinges can get round this problem by having a removable hinge pin, but alignment is trickier. It used to be a common practice to drill through the tab of the hinge and pin it with a piece of cocktail stick, but this is unnecessary unless the hinge is constantly being knocked out, on landing for instance. Most hinges have tabs with holes in them and the glue tends to settle in these and connect the two halves of the post together as a matter of course. Epoxy is particularly good at this.

Mylar hinges consist of strips of Mylar tape, which has an excellent resistance to fatigue fracture and is most useful on surfaces that have a limited range of angles of deflection, such as elevators and to a lesser extent ailerons. They are much simpler to instal, only requiring a slot cut with a knife, with no "V" being required for clearance. The marking out must be a bit more accurate, however. It is a good idea to put holes in each side much as on a pin hinge to allow for glue bonding. The result is a hinge that is springy but has little resistance over small angles. Put in as many hinges as you would for ordinary hinges. Slow setting cyanoacrylate can be used for the installation.

Fig. 2.12

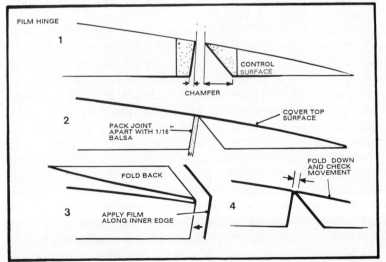

Covering material hinges are made from film covering, and they are particularly suited to elevators and ailerons. The decision must be made as to whether to top or bottom hinge the surface, and the control surface is then chamfered to bring one side to a sharp edge. A thin chamfer is then taken on the opposite side to the main chamfer on both the surface and the spar. Consult figure 2.12 for clarification. The surfaces and the rest of the model are then covered. A strip of covering material is then prepared and, with the surface in line, it is ironed on. The surface is folded back on itself and a second slightly broader strip applied with the iron. The surface is now hinged. If it is stiff peel the hinge off and start again but allow a small gap (1 to 1.5mm) when putting the first strip on.

THE WINGS

Every part of a model matters but the wings have to be right or the whole model is useless. Since the aim of this book is to put you on the road to success it is necessary to skip the aerodynamic theory and just concentrate on the nuts and bolts end of wing building. Two major methods are common at the time of writing: foam and built-up wings.

Foam wings are perhaps the easier of the two to make although a properly stressed foam wing can take as long to build as a built-up wing. The first thing that must be understood about foam wings is that the foam should only be glued with certain adhesives – epoxy, PVA or water soluble latex glues such as Copydex. These are by no means the only ones but they are perfectly safe. If you have something else in mind try it on an offcut first. The instructions with the vast majority of kits are sufficient for the production of a good wing but many foam wings still fail on tow due to excessive increases in stress just at the end of the wing spars or joiner blocks. The answer to this is to run a ply web along the block and beyond its end so that the stress point is pushed further out along the wing and into a less stressed area. Polyhedral wings can have a disturbing habit of crushing the top surface at the root of the tip panel in a minor crash, but if the ply web is run out beyond the polyhedral join it can prevent this happening in all but the worst cases.

The accuracy of a built-up wing and its performance depend very much on your skill on the building board. It is no good trying to build without a building board as the results will not be predictable. Follow the kit instructions and all should be well. If

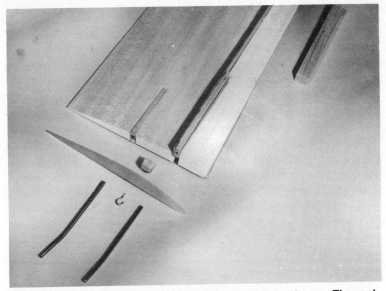

Typical joiner system for a foam wing. Note the full-depth spar. The root rib has yet to be drilled to pass the metal dowels.

washout is called for make sure that each panel has the same amount; either use the same wedge or make both together as a block, then split it to ensure that both are identical. Beware of twists in the building board and go over it with a spirit level before starting just to see if it has moved, as one of mine did once. The wing should have two halves that are matched up to each other as perfectly as possible, since if not the model will turn better in one direction than the other. They usually do anyway, but it should be avoided if possible. If the wing does come out warped steam it over a kettle until it "softens", twist it the other way gently till it stays straight, then pin it down on the building board and let it settle for 24 hours. If this does not work then you are in trouble.

The sheeting is best made into a single wing skin before it is pinned onto the board or glued to the top surface. This is done with the long straight edge. Cut both edges with a sharp blade held at 90° and then tape them together. Fold the joint back, smear glue down the joint, remove any excess, fold back flat and leave the sheets weighted down on a flat surface, with a sheet of polythene underneath, to dry. Build the whole wing skin up in this fashion. Any scarf joints needed to produce longer sheets for panels of over 36 ins. (or a metre) should be done first using the same method. A scarf joint is made by butt-jointing two pieces of wood with the ends of the joint cut at an angle, 45° is adequate,

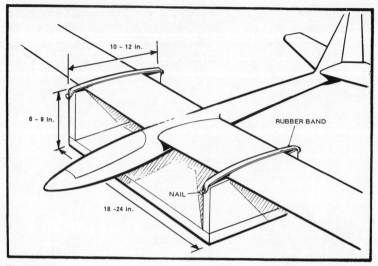

Fig. 2.13

60° or 75° will give a stronger result but is more fiddly and wasteful. The wing skins should be sanded before being used, to help to prevent the positions of the ribs showing, which is likely to happen if the wing is sanded after completion.

When choosing the wing sheeting or wing rib wood select the lighter wood for the wing tip end of the wing and the heavier, stronger wood for the root end. This will help to keep the wing tips light and improve the model's turning response.

As regards adhesives, PVA is adequate for all the rib and sheeting work. Epoxy is better on highly stressed areas such as wing joiners. If time or your patience is short cyanoacrylate can be used for the preparation of the wing skins, positioning of the ribs and spars on the bottom sheeting, the rib cappings and the webbings between the ribs.

On models that use two joiners mounted either side of the spar a considerable increase in strength can be obtained by extending the joiners into a further rib and using a doubler on this rib between the joiners and the spar. A little extra strengthening of the spar on either side of the last "joiner rib" will also help matters as it is a high stress area with all the wing loads transferring into the ribs at this point. It's a frequent area for breakages.

The wing is not just subjected to bending stresses, it is also twisted, or to be technical about it, put in torsion. Torsional stiffness will determine how fast the wing can be flown, though it is a problem that should have been overcome by the model's designer. The root area out to half span is the most crucial area and should be the area strengthened most if flutter or tuck-under problems are encountered. A fully sheeted structure helps greatly, as does the use of tissue, glass-cloth and to a lesser extent heat shrink fabric. Solarfilm does not add a great deal of torsional strength.

If the model has plug-on wings there can be a problem if the wings are not set to the same angle of incidence. It is easy to set one wing so that it gives more lift than the other. Look at the model from the front from a few feet away and see if more of the underside is visible on one side. If it is then the effect is like having ailerons that are out of trim and the model will try to turn one way all the time. This can be held on rudder at one airspeed but will require a different trim setting if the airspeed is changed. The model will not be entirely unflyable but it will not be possible to spot lift easily and the model will be awkward to fly. The answer to this is to use a jig as in figure 2.13. The baseboard must be a stable piece of wood and the support ends must be planed flat and brought to the same setting by the use of a spirit level. The model's wings are built completely and the main joiner installed in the fuselage so that the wings are square with the tailplane. Pieces of "Blu-Tack" or chewing gum are stuck on the fuselage wing roots in the areas that the other joiners are to be placed and the model is assembled and placed on the jig. The fuselage is lined up and supported and the wings are banded down and forced onto the wing roots. The wings are then eased away from the fuselage and the "Blu-Tack" will carry marks where the joiner holes should be. Drill out the holes, instal the tubes and try the model on the jig again to check alignment before glueing it up. Models with banded-on wings rarely suffer problems.

COVERING

There are several methods, but the most often used these days are heat-shrink plastic or fabric, glass-cloth and tissue. Tissue is excellent as a lightweight covering but on open structures it is vulnerable to accidental damage. Unless you are a licentiate to the arcane mysteries of producing a blemish-free tissue finish I would suggest that you use a heat-shrink method. This is not to denigrate tissue, which done properly is structurally much stronger, but since the advent of heat-shrink materials it has become much easier and faster to cover a model and as a result tissue covering has become something of a rarity. Most importantly, the possibility of producing a workmanlike finish without introducing warps is greatly increased with heat-shrink materials. For these reasons I am not dealing with tissue covering in this book. Anyone with free-flight experience will not need to be told anyway.

Slope soaring poses additional problems such as the covering becoming punctured by the sort of hardy vegetation to be found on top of slopes and the frequent rocks that strew hillsides. The result is that nylon covering, because of its rugged nature, is still popular for slopers, making wings less susceptible to damage and fuselages less prone to bursting. The skill is difficult to learn but Keith Thomas strongly recommends the method.

Covering with heat-shrink materials is a relatively simple process. Start by preparing the airframe, sand it properly and then remove any dust with a vacuum cleaner. Fabrics may require a coating of some surface preparation or other on the bonding surfaces to assist adhesion. Plastic films do not normally require anything. The sheet of material is then laid on the area to be covered and either marked and then cut round or just cut round in situ, leave about an inch overlap to aid handling later. Mistake number one is to cut away all the edging on plastic films, which makes it difficult to remove the backing. Do it and you will soon see the problem. If this happens put a small nick across one corner and tear the piece off slowly, then if you are lucky the covering will tear first leaving a piece of backing for you to work on. Alternatively think ahead and peel back the protective film a little, smooth it down again and then cut to shape; the backing can then be lifted with a knife point. The next step is to put the material over the area to be covered and tack it down with a small "travelling iron" or a special "tacking" tool. Two problems can now occur, overheating and bad wrinkles. The bad wrinkles happen if the material is not put on smoothly and probably the easiest way to avoid this is to start tacking on both sides at the centre of the piece and work out slowly towards the ends, tacking both sides as you go. The shrinkage will take out quite a lot of slack but there is a limit and it is best to try to get the material as tight as possible before shrinking. If the result is really bad start again.

Overheating of the material is very easy to achieve and it is a good idea to try a few sample pieces first to get used to the temperature required and the length of time the iron should be in contact. Underheating produces a weak bond with the surface, overheating produces a strong bond but will melt the plastic or scorch the fabric's adhesive. Try to get both effects on your test pieces and then set the iron in between. With the material "tacked" the next job is to iron down the edges all round by about half an inch, which should be done gently so that you do not bruise the underlying structure. Then shrink the material to tighten it. Shrinking should be done when the whole wing, etc. has been covered, and the tools for the job are the heat gun, iron or a gas or electric fire. The heat gun is a very useful tool which like an overpowered hairdryer blasts out a furnace-like breath of hot air (very useful on frozen pipes). Care is required or it can burn through plastic and scorch fabric, so the gun must be kept moving continuously and not allowed to dwell for long on any one spot. The distance from the covering determines the heat build-up. The iron can be used to provide area warming and will do the job quite well, but care must be taken to avoid bruising the structure. A fire can be used, in extremes, but it is a bit hit and miss and it is necessary to catch the shrinkage quickly or the heat may get too much and ruin the job. The material will stick to the surface better if you run over it with a clean handkerchief just after it has been shrunk.

On wings and tailplanes it is best to cover the bottom first and cut the material flush to the edges. The top covering should then overlap the bottom by some 3 to 5 mm. On fuselages do the top and bottom first and the sides last. When shrinking, work along both sides in alternate goes, i.e. do the first 9 ins. of the bottom then the first 9 ins. of the top surface, followed by the next 9 ins. of the bottom and so forth. This helps to avoid warps being intro-duced into the structure. If any warps do appear, bend them the opposite way and reshrink the covering, repeating if necessary until the wing is true.

Glass-cloth is not likely to be used by a beginner or an inter-mediate standard soarer mainly because the extra strength it confers is not entirely necessary for most conditions and models. The beauty of glass-cloth covering is that it substantially in-creases the torsional (twisting) stiffness of the wing, waterproofs it and provides a consistent fairly dent-resistant finish. It can only be fully exploited on fully sheeted surfaces and is best suited to fast models and those that require maximum strength in order to be competitive in competition work. The working method is similar to that used when covering sheeted surfaces with tissue, the surface first being prepared to a good surface finish and dusted with a vacuum cleaner. Any holes should be covered with pieces of masking tape so that they do not get clogged with resin, picked off when the wing is finished. The covering is cut to the shape of the surface with a half inch or so overlap; fraying can be avoided by taping over the lines for the cuts and cutting through the tape. The glass-cloth is then draped over the bottom of the wing with the wing rested on blocks and the table surface protected from spillage. The resin is then brushed onto the cloth, starting at the centre and working out, until the cloth is "wetted out" and the edges are drooping down. The weave should be visible over the whole surface. Roll off any shiny spots with a roll of kitchen towel. Allow to dry and then do the other side. Trim off any excess cloth and cut back to a smooth finish with wet and dry papers, being very careful not to go through the covering.

The resin to use is epoxy, although I have had success with epoxy paint. It can be bought in a viscosity suitable for the purpose. Dope can be used for local reinforcements with glass-cloth, all-moving tailplanes and polyhedral joints being good examples. Epoxy has the advantage of not affecting foam wings, but dope will eat straight through them, so beware of this. Polyester resin is not so useful for glass-clothing, it cannot be used on a foam wing anyway, and it can be brittle and heavy.

For most of our purposes heat-shrink plastic and fabric are of most use and the most acceptable domestically. There are a few more advanced methods for applying the cloth but they need not concern us here.

Finishing a glass-fibre fuselage is entirely different from the usual method employed on wooden fuselages. If it is self coloured it may be left in its moulded colour. If so desired it can be coloured with cellulose car paint which gives plenty of colour choice. The fuselage should first have any bubbles in the moulding excavated and filled with body filler. The whole thing should then be rubbed down until a satin finish is achieved over the whole fuselage. This can then be sprayed with the appropriate primer, followed by the colour coats. Use as little paint as possible as it is all dead weight. Cut the paint back with a cutting compound and apply any transfers before waxing and polishing it with a silicon-free car or furniture wax.

BALLAST STOWAGE

Structurally speaking the best place for ballast is in tubes in the wings. This poses few problems in construction and may well be detailed in the instructions. The ballast tubes should be just in

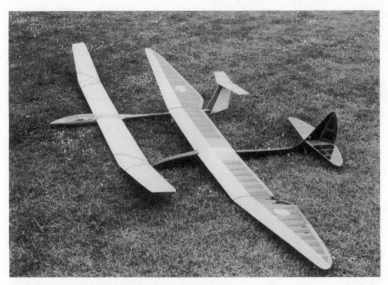

For a second model one of this pair (a *Sunshine* **100in. and a** *Bird of Time***) would be an excellent choice.**

front of the C.G. or positioned either side of it, not behind. It is far easier to instal ballast tubes at the initial construction stage than trying to retro-fit them.

Ballast can be carried in the fuselage between the wing roots, but the space is limited and the quantity of lead that can be accommodated is often limited by the control runs and so forth. Unless the fuselage is designed for such an installation it is perhaps best to stick to loading in the wings.

COLOUR SCHEMES

Due to the sort of heights and distances that we fly our models at we need a highly visible colour scheme. Since the model's colour fades into a shade at any kind of distance the actual colour is not too crucial. However, dark shades and colours are far more visible against a grey sky than a light colour. What we need is contrast and a dark covering will give the maximum contrast. Any colour scheme will benefit from having the undersurfaces of the model in a dark colour. The top surfaces may use lighter shades but it is wise to keep at least 50% of the area in a dark colour so that the top surface is visible when the model is banked over and turning against a grey background. It is a good idea to put a strip of kitchen foil about 1 in. by 3 ins. long on the leading edge on one wing tip, as its flashing on a sunny day at height can help to maintain contact with the model.

Two colours that should be avoided for anything other than small areas on a model are metallic silver and gold. They look superb but fade right out in the sky as they reflect the dominant sky colours, particularly on poor visibility days. Natural finish glass-clothing can also be bad in this respect and may need some colour bands applied to help the model to show up. White gives similar problems.

One not very obvious concern is repair to the covering, not because it is difficult (simply use the processes mentioned above) but because a trainer is very likely to need regular repairs. A simple one-colour covering is much easier to put back in order after a visit to the workshop. One mistake that some beginners make is to do too good a finishing job, with the result that they are frightened of flying the model in case it gets hurt. It is no use having a "precious" attitude to any model: a flying model is only fulfilling its raison d'etre when it is being flown. So for the first few models curb your artistic instincts and do a workmanlike serviceable finish that will not depress you when you have to cut it away to expose airframe damage. Save the masterpiece till you are confident that you are going to be able to keep the model in one piece for a while.

Building well takes a fair amount of experience, so do not be dismayed if your first efforts look a bit rough. By the time that your first trainer has taught you how to fly (and repair) it will look pretty rough anyway.

3   REPAIRS

This chapter is not intended to induce despondency, merely to recognise that the average model will probably have as much time spent in repairing it as it took to build it. This is in the nature of the game and must be accepted. Forget about how well you flew during a flying session, if you return home without any repairs to do then it was a good day's flying!

FUSELAGES

Once broken a vacuum-formed fuselage can usually be repaired by the use of a slow setting epoxy adhesive and a suitable reinforcement material. It can be difficult to effect a discrete repair, though.

With moulded plastic fuselages breakages in the "greasy" plastic used usually come as splits in areas that have been repeatedly abused, e.g. canopy opening corners, which are a favourite. Adhesives for such plastics do not seem to be commonly available, so repairs are perhaps best effected by plating across the broken area with a piece of thin aluminium secured with bolts. Putting electrical tape over the split will help prevent it from getting any worse.

Wooden fuselages are more difficult. Small splits can usually be dealt with by a dab of cyanoacrylate, but larger breakages may require additional stiffening in terms of a piece of ply or the replacement of some of the wood. Areas that are frequently breaking can sometimes benefit from having 1 ounce glass cloth doped onto them; tail ends are a common case. Large splits or springing apart can be dealt with by glueing together and applying, in the case of equipment bays, a layer of glass cloth on the inside, stuck with epoxy.

Glassfibre fuselage repairs are very much dependent upon the degree of the damage. If the whole front end has been smashed it is probably not worth the time and effort. A pod and boom can work well in these circumstances, the pod being discarded and the boom and tail used again if they are in reasonable condition. The two most common breakages are compression fractures in the canopy area and snapped or split booms.

A compression fracture will shatter out the resin from the glass fibres and leave a nasty looking split with bits of fibre showing. The first thing to do is to tape the moulding together so that it looks properly aligned. Heavy matting is then resined onto the inside of the moulding over the breaks. Go about an inch either side of the damage or do the whole side if further damage is anticipated. Prepare the bonding surface with acetone. When the resin is cured remove the tape and then break off any loose pieces of moulding, fill the gaps with a suitable body filler (cellulose car filler or epoxy putty), cut off any obvious excess when it is "cheesy" and when cured cut it back with wet and dry papers and repaint.

Boom tubes on pod and boom fuselages can split down their length and if this is not too bad they can be repaired with a long strand of glassfibre. Wrap it round the split area tightly and fix it with resin, using the heat gun to ensure that it wets out properly. If the split is too bad replace the boom. If the boom on a one piece fuselage moulding breaks it is more serious, although you can usually repair it from the outside or the inside, depending upon how much the appearance matters. The outside repair simply involves wrapping the break in some medium weight

Fig. 3.1

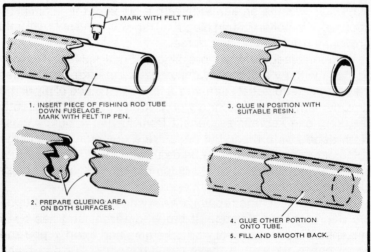

1. INSERT PIECE OF FISHING ROD TUBE DOWN FUSELAGE. MARK WITH FELT TIP PEN.

2. PREPARE GLUEING AREA ON BOTH SURFACES.

3. GLUE IN POSITION WITH SUITABLE RESIN.

4. GLUE OTHER PORTION ONTO TUBE.

5. FILL AND SMOOTH BACK.

glass cloth wetted out with resin. With all g.r.p. repairs remove any paint from the bonding areas and prepare with acetone. Covering the joint with tape as it cures will help to improve the finish and squeeze out excess resin. The repair will then need to be filled, sanded and refinished as before.

Internal repair involves the insertion of a piece of old boom tube (or fishing rod tube) in the forward part of the fuselage and the plugging onto it of the other part. It is fiddly and, as with all repairs to finished fuselages, the control runs must not be glued up. The joint can be disguised with filler and repainted. Fig. 3.1 shows the basic sequence of this type of repair.

Tail repairs almost always add weight and it is a wise precaution to mark the centre of gravity on the side of the model prior to commencing work so that the model can be rebalanced afterwards. Front end repairs will have the same effect, but much less pronounced.

WINGS

Repairs to a foam wing are not as neat as they can be to a built-up wing. A wing that has broken off near the root is probably best replaced completely, or used for a shorter wing. One that has broken off near the tip (or on the tip panel) may be repairable, but the damaged wing will be heavier, and the model will need rebalancing. Slope-soarer wings, being generally shorter, are

Fig. 3.2

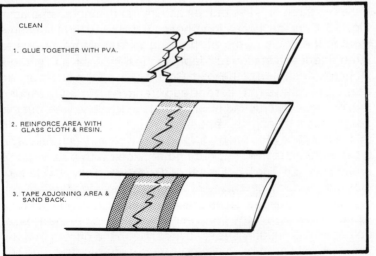

CLEAN

1. GLUE TOGETHER WITH PVA.

2. REINFORCE AREA WITH GLASS CLOTH & RESIN.

3. TAPE ADJOINING AREA & SAND BACK.

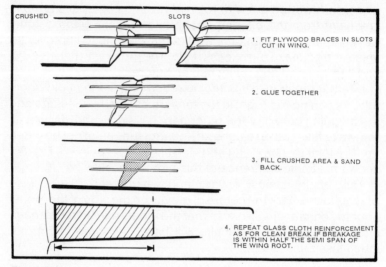

CRUSHED   SLOTS

1. FIT PLYWOOD BRACES IN SLOTS CUT IN WING.

2. GLUE TOGETHER

3. FILL CRUSHED AREA & SAND BACK.

4. REPEAT GLASS CLOTH REINFORCEMENT AS FOR CLEAN BREAK IF BREAKAGE IS WITHIN HALF THE SEMI SPAN OF THE WING ROOT.

Fig. 3.3

more repairable than thermal soarer wings, where weight and efficiency are more important.

Two types of break are possible, clean and crushed. With a clean break glue the two pieces together with PVA and epoxy and if the break is not too obvious glass-cloth the joint with 1 to 4 ounce cloth. Sand back carefully (tape the exposed veneer to prevent cutting into it) and re-cover the damaged area. Fig. 3.2 shows the idea. If crushing has taken place the break may not go back together cleanly, in which case it is necessary to put between three and five braces across the damaged area as in Fig. 3.3. Glue the pieces together as before and then cut some slots in the thicker area of the wing to take some ⅛th balsa strengtheners installed full depth in the joint area, and some 2 inches to each side of the joint. Glue these in, fill in any crushed areas with lightweight filler and epoxy and recover when sanded and dry. A badly crushed break is difficult to repair well, but it is possible to cut back the veneer (balsa veneer is much better in this respect) beyond the break and peel it off very carefully. The crushed foam is then cut out and replaced with a new block, refer to Fig. 3.4. The block is sanded to the wing shape with new glasspaper, and a new piece of veneer butted into the wing. You can then either install some braces across the joined area (to 2 inches either side of the new veneer), or glass-cloth the area, or do both if it is a highly stressed area. Keep the offcuts that the wing was packed in, as they can be very useful as jigs during any

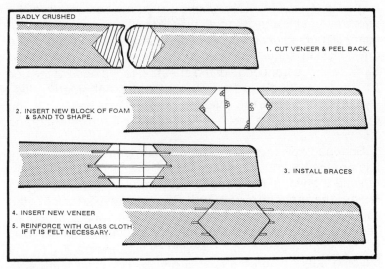

BADLY CRUSHED

1. CUT VENEER & PEEL BACK.

2. INSERT NEW BLOCK OF FOAM & SAND TO SHAPE.

3. INSTALL BRACES

4. INSERT NEW VENEER

5. REINFORCE WITH GLASS CLOTH IF IT IS FELT NECESSARY.

Fig. 3.4

repair process and if used in a carrying box will provide excellent packing.

Built-up wings are generally easier to repair satisfactorily. Breaks come in many different guises but usually involve the following: splitting, crushing, snapping and spar damage. There is a point when the effort involved in repair work is not worth it and if it is felt that it would be quicker to replace the whole wing or panel then there is no point in repairing. If the converse holds true then an attempt is worthwhile.

The first step is to carefully strip the covering from the damaged area. Take great care to inspect for minor damage such as splits along the grain, which may not be obvious. Get a good idea of how much has been destroyed and how much is repairable and plan your sequence of operations. Splits are the easiest to repair, simply run some cyanoacrylate (Superglue) down them and hold them closed. Splits often hide damage to the internal structure, so twist the wing to see if there is breakage inside, maybe even lift the sheeting and have a look before you seal the structure back up again. A split will often creak when the wing is twisted and loss of torsional stiffness is a common sign of splits.

Crushing is not so bad as it looks, but it will involve replacement of parts of the wing. The following is a typical sequence for repairing damage to the leading edge of the wing. Cut out all the crushed parts and leave the ends of the repair area clean and angled for a scarf joint. Now make up rib replacements from the plans or remains and glue onto the stubs of the crushed ribs. Cut some slightly thicker wing sheeting balsa to the shape of the area, leaving some overhang at the front, and glue in place. Sand back the front edge of the sheeting to the ends of the ribs and prepare a leading edge from square material, glued in place with epoxy. Shave the inserted wood down level with the existing wood and to the shape of the wing. Sand carefully and recover.

Snapped wood is best replaced if it is in a stressed area, but if it is not then a spot of cyanoacrylate should be adequate. To replace the wood, cut away at both sides at 45 degrees and replace the snapped piece with a piece of oversized balsa of the same weight and density. Shave down to size, sand and recover. This would be a typical trailing edge or tailplane repair. With all these repairs it pays to pin the wing (or whatever) down on the building board when the joint is drying. The scarf joint, a 45 degree butt joint, is usually stronger than the surrounding wood if executed accurately.

Spar damage is serious. If the spar is intact the wing can usually be saved, but if it is broken or smashed near the root area it is worthwhile considering a new panel or wing, particularly if it is a fast launching model. Beyond half-way out along the wing half (half semi span) the stress is not too bad and new material, scarf jointed with additional reinforcement and glued with slow setting epoxy, may well prove adequate. Tip panels are far less critical stresswise and can usually be repaired if care is taken to ensure that warps do not creep into the construction. Repair it on the building board.

Tailplane repairs are very similar to wing repairs except that weight must be kept to a minimum. Just a small word of warning – do not use cyanoacrylate on foam, it eats into it!

The question of whether to do repairs at the flying site needs some thought. Obviously if you have travelled a fair distance and damaged the model on the first flight you will want to try a quick repair. Beware! Many crashes do more damage than is immediately obvious, site safety must be considered and it is unwise to fly a badly damaged model with field repairs. The main considerations when examining the model must be will it fall apart or lose control in the air? If the repair cannot ensure the control or structural integrity of the model then it should be put away and repaired in the workshop where it can receive proper attention. Flying a previously damaged model is not a good idea anyway, as any further crashes are quite likely to do severe damage to an already weakened structure.

A typical field repair kit could contain the following:–

Adhesives   – Fast-setting epoxy
                 Cyanoacrylate

Tapes   – Electrical
                 Reinforced parcel tape

Tools   – Craft knife
                 Pliers

Materials   – Ply offcuts
                 Balsa offcuts
                 Lollipop sticks
                 Mixing palette and applicators

Spares   – Clevises
                 Wing wires
                 Rubber bands

If the job demands more than these then it is probably unwise to attempt it at the field. Personally I always prefer to repair

A straightforward design is this *Zephyr* **by David Rye, available from Radio Modeller Plans Service Repair of such a structure offers few problems.**

models in the comfort and calm of the workshop.

When inspecting a crashed model pay particular attention to the following: the servo mountings, which often come loose in an impact and flying with loose servos is not amusing. The tail boom/rear end of the fuselage can get whipped quite viciously and become loose due to splits, the tail surfaces' mechanisms and joints can be put under extreme stress and are likely to have suffered some damage (T-tails are usually more prone to crash damage) and the wing joiner tubes; if these have suffered pack up immediately and prepare for major surgery.

Splits in wood can usually be dealt with by a dab of cyano-acrylate, as can non-critical lumps of wood that have broken off. Taping up of a split boom tube will help to prevent it splitting further. Large splits in wooden fuselages can be dealt with by using epoxy. Broken rear ends can sometimes be quickly bodged together by epoxying some reinforcement over the break and taping it up.

Fast-setting epoxy will take much longer to cure on a cold day and any wind will make this worse. Dismantle the model and put it in the car while it is curing or it will take forever. Draping a coat over the model to shield it from the wind will help if the car is at the bottom of the hill. Epoxies, while they harden quickly, take half a day or so to cure properly, so leave the model as long as possible before attempting to use it again.

The covering need not be removed entirely for most repairs, 2 or 3 inches to each side is usually enough. Uncover the damage until you are satisfied that it goes no further. The recovering need only involve a neatly placed patch of material and you might even decide to make it into a decorative trim stripe or club badge position.

4 *TRIMMING AND ADJUSTING*

A model will not fly straight off the building board and is likely to require a fair amount of trimming and adjustment to get it to fly perfectly and perform properly. A radio-controlled model is trimmed in an entirely different kind of way from a free-flight model, due to the fact that it must be capable of being flown at a number of different airspeeds, and have certain stability charac-teristics that are undesirable for a free flighter. There are similarities in the procedures involved but R/C design and its results are different.

There are several phases:

STATIC BALANCE

Once the model has been completely finished it should be statically balanced. This involves hanging the assembled model from a convenient overhead point, a rafter is ideal, on some form of cradle. Several suitable types of cradle are shown in Fig. 4.1. Start off by getting the model's fuselage to balance by moving the pivot point back and forth. When it settles in a reasonable manner put the pivot point over the centre line of the fuselage and you will find that one wing will probably hang down. Take some lead shot or sheet and tape it onto the wing tip of the higher wing. Keep altering the ballast until the wings balance. Next move the pivot point until it is on the centre line of the fuselage and over the recommended position for the centre of gravity. Rehang the model and it will very probably be tail-heavy. Remove the battery and add lead to the nose of the model, replace the battery and check the balance. The model should hang either dead flat or slightly nose down when the correct amount of ballast had been added. Check that the pivot point has not been moved during this process and take the model down.

The wing tip ballast should be buried somewhere in the wing tip area, usually in the balsa block that protects the end of the panel. Cut a small area of covering away (you could finish the wing tips last, of course) and excavate a hole for the lead. Fix it in place with epoxy and cover the hole. The nose ballast may be fixed with epoxy if it is in the form of shot, or cut to shape and carefully packed flat against the front bulkhead if it is sheet lead. If the model is nose-heavy put the ballast in the base of the fin (wooden fuselage) or the bottom of the fin post.

Getting these two right will ensure that the centre of gravity is more or less correct and that the model will not tend to drop one wing when it stalls.

CONTROL MOVEMENTS

A very common fault is to get the control surfaces moving the wrong way, which even experienced fliers have been known to do. The first thing to do if you are a newcomer is to get an expert to check the model over prior to the first flight.

The rudder should be set up so that when viewed from the rear it moves to the right when the stick is moved to the right and vice versa. The elevator should point upwards at the trailing edge when the stick is pulled back and downwards when the stick is pushed forward, just like a full size joystick. Fig. 4.2 may clarify this. Ailerons are a little trickier. View the model from the rear, when to turn right the right hand aileron should point up and the left downwards.

Fig. 4.1

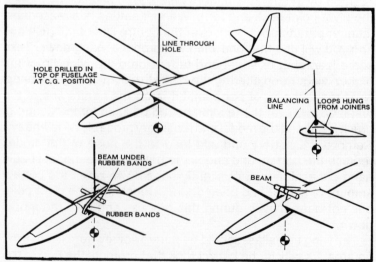

TEST GLIDING

All the initial flights should be carried out by an experienced flier, which prevents the beginner from making a mess of the model immediately and maximises his chances of success later.

Opinion is divided about the value of hand-launching a model; personally I feel that it is the best starting point since you do not have a lot of height to damage the model with if it is drastically out of trim. Pick a day with a steady 5 to 10 mph wind for the first trimming session.

The first hand launch should be used to get the elevator trim roughly right. The rudder is centralised and the model is then run into wind and gently thrown, with the nose pointed about 5 degrees downwards, when it feels that it wants to fly out of your hand. The model will do one of three things – rear up and try to stall, fly straight, or dive. If it rears up full down should be put in until it noses over and regains flying speed, and it will then need to be held in a level flight trim until it lands. If it dives full up will need to be applied until it levels off and can be held level. If it flies level straight away congratulate yourself. The level flight position should be remembered and the control surface position noted. The transmitter trim should then be centred and the clevis adjusted on the connector until the surface is roughly in the level flight position. Any adjustments to servo connections are better undertaken with the gear switched off.

Fig. 4.2

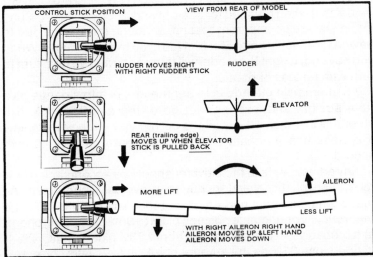

Try a second hand-launch, which this time should be less dramatic and it should be possible to get the model to fly straight and level with the elevator trim on the transmitter. Keep adjusting and hand-launching until it flies level with the transmitter trim in the middle position. The rudder is a far less sensitive control and can be adjusted last, getting the model to fly straight with the transmitter trim in the central position using the same method of hand-glides and clevis adjustments.

The static balancing will have put the centre of gravity (C.G.) in a reasonable position and having got the model flying reasonably well it is time to try a proper launch. Put the model up on the line without using the elevator and do not be bothered if it does not go to full height. Observe the model and adjust the trim until it flies comfortably and is not stalling. What we are looking for now is the correct position for the centre of gravity, which will be slightly different for every model, even ones to the same design. Ease in the up elevator and bring the model gently to the stall. If it rears up badly, stalls and it is difficult to get the model to regain flying speed and recover from the stall then the C.G. is too far back. If it does not want to stall and recovers quickly with a rapid increase in speed then the C.G. is too far forward. The best position is the one that suits you and is comfortable to use, somewhere between these two extremes. When you find it stick with it.

Another sign of a forward C.G. is the flying behaviour of the model. If it flies fast and comes down quickly all the time then it probably is badly out. A fairly certain sign of a rearward C.G. is a model that when trimmed for straight and level is forever getting into a stall for no apparent reason, although this can be due to having too much up elevator trim fed in on the transmitter. When you have got it right the model will fly sweetly and go from turn to turn with no fuss or drama.

An alternative method is to put the model into a gentle dive from straight and level and let go of the elevator stick. If it steepens in the dive the model is tail heavy, if it recovers rapidly it is nose heavy. Ideally it should pull out gently of its own accord.

Any adjustments to the elevator should be accompanied by a recentring of the elevator trim as previously mentioned. All these empirical (trial and error) methods will put the C.G. in a theoretically predictable position of 25 to 35% of the wing chord, and probably in a narrow band of about 33% plus or minus 2%. If in doubt use 33% as the starting point – it is usually correct.

CONTROL THROW ADJUSTMENTS

By now the model should be flying in a respectable manner and be predictable in its behaviour, so the time has come to tailor its control responses to your preferences. Make only one adjustment at a time so that its effects can be easily gauged.

The rudder is not usually too much of a problem. It should be set up so that it gives a rapid response to control movements, but if it does not do so then three courses of action are open to you. First try an increase in rudder movement; up to 30 degrees each side can be used, but anything over 30 degrees is a waste of time as the surface is likely to be stalled when the rudder is moved beyond this limit (you can try this effect out in the kitchen sink with a broad knife). Secondly you could try an increase in dihedral, which can usually be achieved by bending the joining wires up a little at a time. Be careful though, for extra dihedral will make the model easier to fly in a circle but more difficult to fly in a straight line, particularly at speed. Find a dihedral setting that is a comfortable compromise for you. Lastly, if the rudder is set for 30 degree movement and the handling is to your satisfaction but there is still not enough control response, try extending the top of the fin and rudder. This increases its aspect ratio and efficiency, and experience has shown that increasing the width of the rudder is far less effective. Any increase in the fin and rudder area may require some adjustment to the dihedral.

If the rudder is far too responsive the easiest thing to do is to reduce its movement. It may well be worthwhile to reduce the dihedral too, if it looks excessive.

The elevator is another matter, especially as the two types pose different problems. The standard elevator is usually less troublesome and gentle in its operation. It should be set up to give about 15 degrees of movement each way, though this is only a starting point. It will quickly become apparent if the model is oversensitive, because it will be very tricky to fly and hard to settle after a stall, no fun at all. The only thing to do is to reduce the movement until the model is comfortable to fly. Ideally it should be capable of performing a reasonable sized loop on full up elevator. Hand-launch gliders benefit from having a more powerful down elevator action so that they can quickly level off at the top of the launch.

All-moving tailplanes (A.M.T.) should be treated in the same way. Set them up with an initial control throw of some 7 or 8 degrees of movement each way and reduce it if necessary. The majority of all-moving tailplanes are usually built with far too much throw on them. This is due to the geometry of the linkages, so always use the bottom hole on the A.M.T. control horn and the inside hole on the servo arm as a starting point.

The converse case of an insensitive elevator response can usually be tackled by simply increasing the movement. Move the clevis to an outer hole on the servo.

The centre of gravity (C.G.) position will affect the elevator response. A rearward C.G. will make the elevator more sensitive due to the model being less stable. A forward C.G. will make the up elevator response sluggish and the down response rapid, due this time to the high degree of stability produced. This can be used as a trimming method too, aiming for a compromise position.

When you have come to a satisfactory trim state it is worthwhile, when a standard elevator is fitted, to adjust the angle of attack of the wing or tailplane so that the elevator lies in line with the tailplane at the normal straight and level trim position. This will help the elevator to respond evenly.

SERVO ADJUSTMENTS

Wherever possible make use of the outer holes on the servo output arms, which gives greater definition to the control command and reduces slop in the control run. The servo arms should be lined up directly across the fuselage, at 90 degrees to

It happens to everyone. A *Proton* about to bite the dust – it was repaired to fly again!

the centre line, when in the straight and level flying position, see Fig. 2.4. This will ensure that the surface gives an equal amount of movement each way. On the other hand you may want to offset the arms to give a differential movement, and although this is satisfactory up to a point, it is better to engineer differential movement into the linkage and leave the arm at 90 degrees. Linear servos should be set up with the outputs level with each other.

TRANSMITTER ADJUSTMENTS

Ideally the trims should be centralised when the model is in its straight and level trim. This then allows for up trim and rudder trim when thermalling, down trim when penetrating in a wind and fine adjustments in between.

Rate switches are a mixed blessing. They can be used to adjust the control response of a sensitive model by reducing the control throw, but in so doing they will limit the range of movements available to the servo. This will make the control movements coarse and lacking in fine definition. After the initial trimming session adjust the positions of the clevises on the servos (and the connectors on the surfaces if necessary) so that the servos give the movements found during trimming when they are set for full movement. This adjustment will recover the full definition of the servo.

The rates should then be adjusted so that the model is very docile when they are switched in. This is a good position to have them in for learning, and for thermalling in marginal conditions when the extra drag of large control movements will bring you down faster. The only problem is that there are times when sudden evasive action is called for, and at these times they should be switched out immediately. Landing on a windy day, or on a flying field liable to turbulence, is best carried out with the rates switched out. Rates are very worthwhile but they are something else to think about and go wrong, so learn to fly without them if you can.

TOWHOOK ADJUSTMENTS

Undertake these last, as the position is dependent to a certain extent upon the C.G. location. The first step is to observe the behaviour of the model on the line. It will do one of three things to a greater or lesser extent.

If it tends to have its nose pulled down by the line and require up elevator to achieve full launch height then the towhook is too far forward. Launches in calm weather will be poor. This is, however, a very safe set-up since the model is less likely to stall on the line since tension is being wasted to produce excess airspeed. For training purposes a model that has a slightly forward towhook is a good idea. A lightweight model may also benefit from using this position.

The opposite case is far more tricky to fly. If the model launches very easily but requires down elevator to prevent it from stalling then its towhook is too far back. Under a lot of line tension the model will veer from side to side in a barely controllable manner and may stall on the line at low altitude. This is not acceptable and the hook should be moved forward.

The ideal case is where the model launches to full height without the trim, control stick or elevator being altered from its straight and level setting. When this is found stick with it and use it for all reasonable windspeeds. Higher windspeeds can be handled by the application of down trim or down stick.

The towhook position determines the load on the tailplane, as the wing has to do the work in any case. A forward hook position puts it under strain as it tries to hold the nose up, the reverse case applies for the rearward position.

TURBULATION

If, after all the above, and a check that the structure is accurately built, the model still does not perform as expected it may be due to the low loading of a fast wing section. This can happen if the model is covered with a high gloss covering such as Solarfilm. The symptoms will be that the model does not perform well unless flown fast, while other examples of the model covered in a textured covering like a heat-shrink fabric are happy at low speed. The possible answer to this is to turbulate the wing. Place a strip of 3mm car trim tape along the wing at the 20 to 25% chord position, which may help the air's transition from laminar to turbulent flow and improve the low speed performance of the model.

What do you get for all this trouble? A model that is easy to fly and will make the learning process as painless as possible. Failure to trim the model properly will lengthen the time that it takes a beginner to become proficient enough to fly solo, and make learning to fly in thermals a much more difficult taks. It will take time to sort the model out properly but it is time well spent. Some models are only at their best after a season's flying and familiarisation. Keep at it till it comes right.

5   LAUNCHING

The purpose of the next three chapters is to run through the stages of typical flights and give the beginner some idea of the problems likely to be encountered. After much thought I felt that the actual process of teaching people to fly was too diverse to be analysed. People come to soaring with such different experiences and abilities that it is difficult to know where to start. Instructors work in different ways and to prescribe a set method would be confusing. It is hoped that these notes can be worked into a beginner's learning schedule to aid in the process of learning to fly. These chapters may help to avoid disasters or to explain them. In addition it is hoped that these flying notes might help those with some experience to overcome or understand some aspects of flying that may be causing them trouble.

Launching is a logical place to start with from a writing point of view, but if you are an absolute beginner do not attempt to launch your model until you can fly with some confidence. The best sequence for learning is flying first, landing second and launching last; complete solo flying follows shortly after. Read chapters 6 & 7 first and then come back to this one.

BUNGEE OR HI-START LAUNCHING

This style of launch is best suited to the beginner and general sport flyer, and it can be achieved single-handed. There are three basic types of bungee as already outlined in chapter 1: surgical, cotton-covered and solid rubber. Each one derives its name from the elastic member used to provide the motive power. The bungee (pronounced bun jee) is effectively a single element catapult comprising a length of elastic material, a length of fishing line, a parachute, ring fittings for tethering and connection to the model, a ground tether and finally a reel of some sort for storage and retrieval.

Surgical tubing bungees. These are the most expensive and the most useful for a range of conditions and models. The elastic

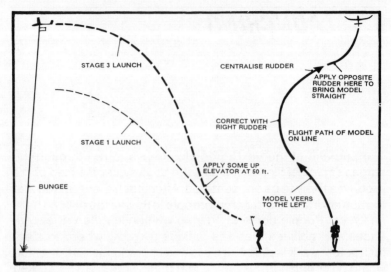

Fig. 5.1

member is some thirty yards of surgical tubing, usually imported from America. The reason that these are better is that they stretch further and pull more gently. The "rubber" being in tube form stretches further because of the hole; solid rubber reaches its limit earlier because it cannot compress the rubber inside the strand. This extra stretch means that they will require a slightly larger field. They are particularly good in moderate conditions with windspeeds in the 5 to 10 mph range.

Cotton-covered bungees. The elastic member here is a bundle of strands bound in a cloth covering. They are usually cheaper, and because the cloth covering limits the stretch and protects the rubber from abrasion, they usually last much longer. The material should be familiar from its use on baggage straps. These will not give such a good launch height in calmer conditions and tend to be fiercer in the way that they release their stored energy. They are more suited to stronger models and to a certain extent heavier models that require a faster launch. The rapid acceleration they produce quickly pulls the model through the critical first 50ft. of the launch, thus avoiding possible stalling problems. In a wind they are as good as a surgical bungee.

Solid rubber bungees. These have been experimented with but have not been too successful with larger models. The stretch available seems limited in comparison and the release rather fierce, so the above two are to be preferred. However, if you have a ready supply of suitable rubber give it a try, nothing ventured nothing gained. A mini bungee can be used on hand-launch gliders for those of you who do not feel up to the athletic skills of glider throwing. For this use a piece of ⅛in. strip free-flight rubber some twenty to thirty feet long and make the overall length some 100ft. or so. Small sites that are close to you become possible flying areas with such equipment.

TECHNIQUES – LEARNING

The first thing that must be learnt is not to be intimidated by the initial surge of power that a bungee produces, and the drama of the first 100ft. of height. It's always the part that impresses small boys! A rapid acceleration and climb out are much safer than a slow climb out with possible stalling problems.

Rule Number One – Use plenty of stretch, particularly in calm conditions.

The next point to beware of is the increase in control response that the launch produces. The pull of the line will produce an increase in dihedral, which will make the rudder twitchy and can lead to an erratic climb. The elevator will be more sensitive, although this will depend upon the towhook position (chapter 4).

Rule Number Two – Do not use large control movements on the line till you know the model.

When you are first learning to launch try the following sequence with a well tried and sorted model:
STAGE 1. Trim for straight and level flight, stretch the bungee and launch with the nose of the model held up at 15 degrees. Only touch the controls if the model is going badly off line. See Fig. 5.1. The model will climb out to some 300 to 400ft. without any undue fuss.
STAGE 2. When stage 1 feels comfortable try steering the model on the line with small movements of the rudder, Fig. 5.1 again. You will find that it is necessary to put correcting movements in. Steer right, centralise the stick, model will continue to go right, put some left stick in and the model will point back to the vertical. The model may now be climbing

How an expert
does it. Al Wisher
launches his well-
known *Blue Beast*
model.

slightly crosswind and may want to come left to get back in line
with the wind. Let it come left and give a dab of right stick when it
is back into wind to make it climb "vertically". Get used to this
sort of thing above 100ft. in case you overdo it and need height
to recover. Try to apply course corrections early, remembering

Fig. 5.2

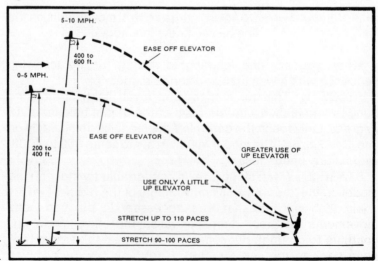

that the pull of the line may well tend to exaggerate the turn.

STAGE 3.   When you feel that you have mastered steering on the line try to ease in some up elevator on the climb (Fig. 5.1). This will increase the lift that the wing produces and slow down the release of energy that the bungee rubber holds. This slower release combined with the effect of kiting up against the wind will give a higher launch. Start to ease the up in at about 100ft. to start with and watch out for the stall. If the model stalls on the line it will very probably tip stall and veer off to one side, which is where the stage 2 work comes in. Apply full opposite rudder to try to straighten the model up and ease off the elevator so that the bungee can give it some air speed. As the model accelerates it will suddenly turn back, so be ready to give a correction as soon as it approaches the correct heading and watch out for over-correction. Ease the elevator back in and keep going.

TECHNIQUES – ADVANCED

As you get to know the model you will learn how much up elevator can be applied on launch. You can now start to apply up elevator earlier in the launch, bring it in earlier and earlier until you can grab for it as soon as the model is thrown. This is the secret of a high launch with a bungee – the sooner the model is pulling hard back against the pull the better. The energy stored in the bungee will then be released at the slowest possible rate and the maximum height attained. You must know the model well to get away with this.

On a windy day the model will "kite" up, that is to say that the wind has enough energy to lift the model by itself. The stretch-ability of the line can then be used to give a higher launch. The trick is to fly with as much up in as possible and fly gently from side to side. This stretches the bungee to its limit for the model and windspeed. When the model reaches the top it can either do a "ping" (more on that later) or if fitted with a releasable towhook the line can be simply released. The extra height that this can gain can be dramatic.

On a calm day the model will benefit from being launched fast like a javelin, which will immediately put it at flying speed and save the energy that would otherwise be used up in acceleration.

As the flying session progresses the wind may swing and the line drop in an awkward position. Until the line is relaid it is worthwhile carrying it crosswind before releasing it so that it drops back in the original position.

WINDSPEED EFFECTS

0 – 5 mph Bungees are disappointing in calm weather, the stretch soon dissipating as the model climbs; a bungee really needs the assistance of the wind. There is little point in trying to use a lot of up elevator, as the model needs airspeed and is getting it from the bungee and excess elevator will only stall the model, see Fig. 5.2. Accept a low launch. The bungee should be stretched to close to maximum length, roughly 110 to 120 paces for surgical type.

5 – 10 mph All the techniques discussed so far can be used for this range of windspeeds. This is an ideal windspeed range for a bungee. A well-trimmed model should just about fly itself up the line. Apply up elevator carefully, as up to about 10 mph tip stall is likely. Stretch a surgical bungee to about 100 paces, Fig. 5.2 again.

10 – 15 mph A bungee is pretty safe in a moderate breeze, as its ability to absorb gusts by stretching is most useful. As the wind rises the bungee will be able to handle bigger and heavier models and the model need not be thrown so hard. Up elevator can be applied immediately and the stretching technique used. Use about 90–100 paces of stretch. Fig. 5.3 illustrates.

15 mph plus Only some 80 paces of stretch are required as the wind will stretch the bungee out. Up elevator may be unnecessary and might allow the model to overstretch the line with the danger of breakage. In fact some down elevator may be required. On release the model will shoot up and drift behind you as the line lengthens, which can be disconcerting but is normal. If fitted with a straight towhook the model can be difficult to get off the line and you may have to dive and pull up hard. Fig. 5.3 again.

Over-stretching will shorten the useful life of the bungee and some 120 paces should be regarded as an absolute maximum for a 30yd. piece of surgical tube. You can tell when you have reached the elastic limit of the rubber for both types by feeling

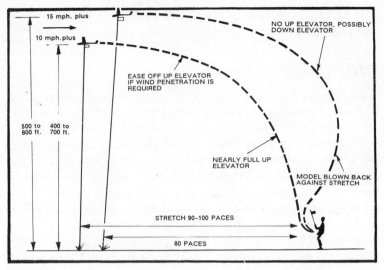

Fig. 5.3

the tension in the line. It will increase steadily and then go very stiff. It can be stretched further but it is only the nylon fishing line stretching. The point to stop at is when the line goes stiff, when you should slacken it off by some 5 to 10 paces and regard that as the limit for the particular bungee in use.

Repairs to a surgical bungee should be carried out by the insertion of a piece of dowelling some 2in. to either side of the break. As the rubber is stretched it will grip the wood tighter and tighter and no adhesive is required. Knots are to be avoided.

TOWLINE LAUNCHING (HAND TOWING)

This is probably the best method if performed properly. The equipment, the winch, consists of a bench grinder gearbox fitted with a reel and lead-out for the line, a handle and 150m of monofilament fishing line. The line used is determined by the usage; 35 to 40lb. breaking strain line is adequate for sport use, 60lb. line is needed for the average competition model and 90 to 100lb. is necessary for the modern fast launching breed of model.

Hand-towing is a two-man affair and requires co-operation and understanding from both parties. Good piloting can make life easy for the towman and the towman can do much to save the model from undue strain. In competition this co-operation must be well practised. The towman's role is as crucial as the pilot's. You must sort out a set of signals for the following techniques.

PILOT TECHNIQUES

In many ways it is much like bungee launching, except that once the model is established in the climb the elevator should be held steady and the line tension controlled by the towman.

To launch the model you first give your start signal (a raised leg, waved transmitter or model will do). You then wait as the towman runs and let the tension rise in the line, until when it reaches that expected of a good bungee you release or throw the model into the air at a 30° angle. Let it accelerate, put some up in as soon as possible and settle the model in the climb. As it reaches the top put some extra up in to build the tension up for the "ping", or allow the tension to drop and the model to fly off the line.

In calm weather, if you have a good towman, it is possible with the right model to get right to the top of the line, which is a big advantage of this method. The model should be launched in the normal fashion but thrown hard, it is then up to the pilot to feed in as much up as is possible without stalling. This gives the model "bite" and gives the towman something to pull against.

Windy weather should be approached in the usual fashion but the up elevator should not be used and down trim can be fed in high winds. If the model is taking an unacceptable strain feed

Line tension demonstrated by Bob Page and his *Selestra* **design, about to start a competition flight.**

in a little down, slowly –don't bang it in or the model may dive and get tangled up with the line. As with the bungee, less line tension is required prior to launch.

Do not play around with the elevator during the launch, as the towman should govern the line tension. He can release tension fairly fast but cannot build it up very quickly. A dab of down may result in a hectic run to rebuild the line tension. Do your best to keep the tension steady.

At the top of the launch you have a choice of how to release the model from the line; loose without tension, dived and pulled off, or do a "ping". The old method is to let the line tension die and let the breeze blow the line off the model, which does have the advantage of letting the model start its flight at a normal speed. All the towman has to do is let the tension die once the model has reached the top. A more normal method is to dive gently and let the pennant pull the line off, and all the towman has to do here is indicate that the model is overhead.

The ping launch is different. The line tension is kept high or increased towards the top of the launch and the towman has to ensure this. When the towman indicates that a high tension has been achieved the model is dived almost vertically to use the line tension to accelerate it. When the flag falls off the model is immediately put into a 45 to 60° climb and levelled off when it has slowed down to the stall speed. This is a technique for use on strong models only. Tell the towman the sort of launch you require prior to him taking the line out.

If a thermal is encountered during the launch the towman may well be aware of it, so arrange a signal for such an occurrence. Come off tow at the top and come back searching for it.

If the model tip-stalls the procedure is a little different from that used on a bungee. Apply full opposite rudder, leave the elevator alone, and hope that the towman keeps running. If he does keep going the line tension will return and the model will be able to accelerate, regain control effectiveness and flown back onto course. If the model is diving into the ground at a great rate of knots pull in all the up and come off the line. The recovery is usually complicated by the model's airspeed, position and the likely proximity of obstructions downwind, so you may need just a little luck.

To abort a launch that is in progress, simply apply down elevator and fly off the line. If the line breaks you need to react quickly: immediately apply full down elevator to prevent the model stalling, regain flying speed and land.

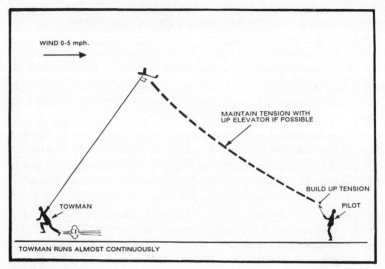

WIND 0-5 mph.

MAINTAIN TENSION WITH
UP ELEVATOR IF POSSIBLE

BUILD UP TENSION

TOWMAN

PILOT

TOWMAN RUNS ALMOST CONTINUOUSLY

Fig. 5.4

TOWMEN'S TECHNIQUES

Model towing is something of a responsibility and requires some skill. Learn towing by launching strong models at first, F3B, 2 Metre or fast Open Class competition models should do. Then work your way down to the lighter models as your finesse improves. In many ways it is like launching a heavy kite.

The key to good towing is managing the line tension, and you only have two indicators, the pull on your hand through the winch and the whistle in the line. Each model will require a different amount of tension and at different points in the launch it may need to be altered. So you will have to learn the launch behaviour of each model that you may need to tow if you are going to be called upon to pull a model up in a contest.

To increase tension you run forward, to decrease it you stand still and let it die off or run back towards the flier. Somewhere between these three positions is a method that gives a steady tension of the right strength, and your job is to find it and maintain it. The pull on your hand that a model requires must be learnt, but the whistle that the line gives out is a fair indication and can tell you a lot. A lightweight model should not cause the line to whistle, and if it does ease off. A medium-weight model can usually take a whistling line, so get the line to just start to whistle. A strong model can be towed with the line whistling all the way. A strong heavyweight model can be torn into the sky

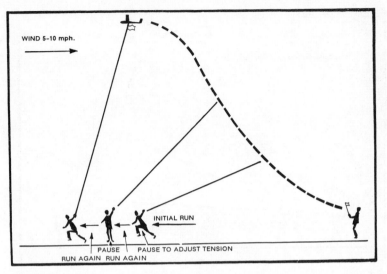

WIND 5-10 mph.

INITIAL RUN

PAUSE | PAUSE TO ADJUST TENSION
RUN AGAIN RUN AGAIN

Fig. 5.5

with the line shrieking! The pilot should say what sort of launch he requires prior to the flight.

The first part of the launch is crucial. When the signal to run is given turn and run forward. The tension will start to slow you down and you may think it too great, but DO NOT STOP. The responsibility for the release tension is the pilot's, so keep going. When the model is released the line will start to go slack and you must speed up to bring it back, as the model is climbing for the first 50ft. and the elevator has not been set for the climb. As the model bites and settles into the climb the tension may rise very rapidly. Stop it rising and let it drop to a reasonable level for the model, then run, wait, or run back towards the pilot to keep the tension at that level. As the model reaches a height where the line is at about 75° to the vertical run hard to generate tension for the "ping", signal that the tension is high enough with a raised arm, and keep going till the model is released from the line. The pilot may not want a "ping" in which case pull the model overhead, signal that it has arrived there and let the tension drop right off.

When the launch is completed reel in the towline so that some tension is present in the line. If you let it drop to the ground the friction should be adequate. Reel in at a steady pace (it is not a race) and check that the line is spooling onto the reel tightly. If it is very loose it is very likely to end up in a tangle, so lay it out again and try again. Then you rest.

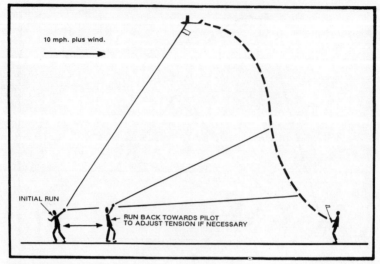

Fig. 5.6

To give you some kind of idea of what to expect the following are typical of what you might experience:

0 – 5 mph	A flat calm is a nightmare, all you can do is sprint flat out and hope that the model bites enough for you to be able to slow down. If it does bite and a respectable pull results do not slow down much, keep it going. As the wind rises the job gets easier and the running less fast. Fig. 5.4 shows a calm weather sort of launch.
5 – 10 mph	Moderate weather is much more comfortable. The model can be played up the line like a fish and you can ease off when the tension is high or run when it goes slack. This is an ideal sort of windspeed to learn towing in. Fig. 5.5.
10 – 15 mph	Towing is easy at these sorts of windspeeds and requires little effort. However, it is easy to over-tow and pull the model apart, so some experience is required and finesse is needed. Be ready to retreat towards the pilot. The run forward at the end will still be needed. Fig. 5.6.
15 mph plus	A very short initial forward run is needed followed usually by a rapid retreat towards the pilot and a short run as the model reaches the top.

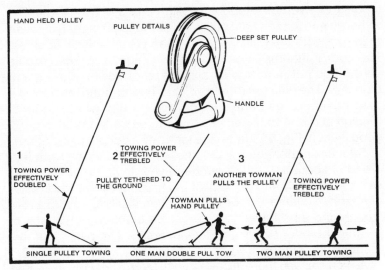

Fig. 5.7

To complicate matters there is likely to be some form of atmospheric action going on, sink or lift. Sink makes life hard for the tower and all he can do is run on through it. Lift on the other hand makes the job very much easier, the model rising quickly on the line with an unexpectedly strong pull. If you have had to run most of the way in previous tows and you find yourself standing still or retreating then you are probably towing through lift. Signal the pilot and continue towing. This signal will need to be seen by the pilot's spotter and arranged prior to the flight.

The towline should of course be laid out into wind and you can throw grass into the air for a guide, but be aware of any windshifts that happen while you are waiting and adjust your position accordingly.

In competition it is a practice to lay out two towlines in case one breaks. Make sure that they do not cross. If one does break, reel it in immediately, pick up the spare line and wait for the pilot to sort himself out. If a line breaks during a sport session watch carefully to see where it lands, then reel in what's left and join the search. If only one line is used at a contest and the pilot makes a hash of the launch do not reel in, let a helper retrieve the line. Walk back to the start position and start again. If the line cannot be found quickly, or if it has broken, a second winch may be run out. Run to get it and get back in position as soon as possible. Two lines are far better in competition.

Pulley towing adds a new dimension to launching heavy

models or average models in calm weather. The towline is tethered and a hand held pulley attached to the line. The tower then runs with the pulley and his speed is effectively doubled. The sort of models or conditions that require the use of this method do not need too much finesse as power is the prime objective. The pulley will feel as if it has far too much tension on it (twice as much as normal at the tower's end gives a normal tension at the model) and you need to be quite brutal to get a good launch. Fig. 5.7 shows the idea and a couple of variations.

Sooner or later you will pull the wings off a model, and feel dreadful about it. It may have been your fault, but if you were working sensibly and to the pilot's instructions then you are not ultimately to blame. It is the pilot's responsibility to ensure a safe launch and as a tower you cannot be blamed for his lack of judgement, poor building or lack of ability. The pilot is always responsible for the safety of his model!

As you can see there is quite a lot to learn about towing, so do not rush it. Towing is a satisfying skill in itself and should be an ability that every fit flier should have.

POWER WINCHES

The power winch is not really a beginner's tool although it is a docile enough piece of equipment if used properly. Safety is a prerequisite as a lot of power is involved and the line, if it touches you when it is in motion, can cause nasty cauterised cuts. Never operate a winch if someone is near the line. Club rules for their operation differ, but stick to them.

For launching the equipment falls into two categories, straight winches and line tensioner winches. The line tensioner winch

The line-tensioning mechanism on a power winch.

Typical of power winch design is this American example. Many clubs and individuals build their own.

has a mechanism that limits the amount of tension that there is in the line to a preset value. This value alters with the weather and the model and the winch operator should be able to set it to an appropriate value. All that one need do is regard it as a towman that does not run out of breath. Step on the foot switch and wait for the tensioner to cut the motor out. Launch and climb out with the winch switching in and out as the tension adjusts, using ordinary hand towing practice. At the top either stop the winch and let the model float off the line, or ping it off by following the usual practice. For a ping the winch should be switched on as the model dives and the motor will cut in; when the model is travelling fast enough stop the winch, let the parachute drop off and climb out on the speed produced. An advanced technique, this, and one that requires a strong model.

A straight winch does not use a tensioner and the foot switch has to be pulsed or blipped to keep the tension correct. For this reason I would recommend some practice on a line tensioner winch first, which will give you an idea of when the winch needs pulsing.

When the launch is completed the winch should be switched off and the non-return ratchet disconnected. The line is retrieved for the next launch, the ratchet reconnected and the line tethered ready. The non-return ratchet prevents the line from spooling out when the winch is not pulling the line. You can

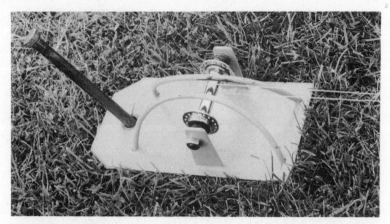

The "turnaround" for a power winch, used when the flier both launches his model and operates the winch.

stretch a bungee in the wind and the line from a winch drum can also be pulled out if there is sufficient pull from the model, hence the ratchet. This has one potentially bad effect in that once tension has been built up in the line it cannot be released by the winch. Do not put any more tension on the line than is absolutely necessary, and watch out for gusts of wind. Use some down elevator if necessary.

6   FLYING

As we leave the towline behind the fun really starts. Thermalling is the subject of another chapter and these pilots' notes will deal with the basic manoeuvres.

THE STALL

The first and probably nastiest thing to master, the stall causes more problems than any other aspect of flying. The sequence of events is usually this: the model flies along and starts to slow down, the nose rises, it rises further and the speed drops off. At this point the model almost stops in the air and the nose drops, pointing down in a shallow dive (a classic stall), and the model accelerates and then starts to climb again. This time its nose rises faster and sharper and it stalls in a more savage manner. If left alone this sequence will develop further. The cause here is the trim of the model, i.e. too much elevator is trimmed in. The first thing to do is to feed some down trim in. The next is stall recovery. Consult figure 6.1.

Stall recovery requires timing and probably the best time to kill the stall is as the model rises up. Watch for it to rise and as it approaches the stalling position give it a healthy dab of down elevator. This should level the model out at something like flying speed and if the down trim has been reset properly it will then fly off normally.

More often than not the model will try to stall again, so if it does give it some down elevator as it rises and watch for the start of another stall. If it continues to behave in this way the elevator is still not right, so feed in a little more down and watch the results. Remember, speed kills the stall, use down trim.

One way to kill a persistent stall is to put the model into a turn. It will probably settle down, and when ready you can fly straight again and tackle the trim problem. This can be useful just after launch, to settle the model.

If the model is balanced with a rearward C.G. it can be difficult to recover from the stall; the opposite is true for a forward C.G.

DIVE RECOVERY

This can require care. You apply up elevator and wait for the model to start to climb, centralise the stick and watch the speed, apply down at the normal flying speed and settle the model down. If you are travelling very quickly put the up in slowly and progressively as a rapid application of elevator may overtress the airframe and pull the model apart. If airbrakes are fitted use them. If any part of the airframe starts to flutter, come out of the dive immediately.

If the model noses over and goes into an ever-steepening dive it has tucked under and full up will just hold it in a vertical dive. Ease off the up elevator and apply down elevator, when the model may do an inverted loop and it can then be brought back to straight and level by half rolling back to an upright position. You need height for this manoeuvre, usually lots of it. Tuck-under is a sign of structural problems in the wing (lack of torsional stiffness), an undersized tailplane or a sloppy elevator linkage, probably a combination of them. Sort it out in the workshop.

STRAIGHT AND LEVEL FLYING

Having killed the stall we are now flying straight and level (S & L).

Fig. 6.1

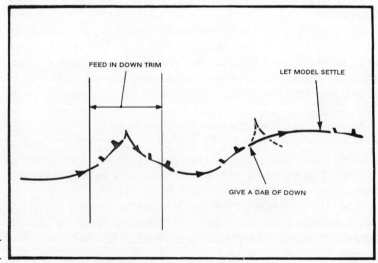

FEED IN DOWN TRIM

LET MODEL SETTLE

GIVE A DAB OF DOWN

Well actually "level" is a misnomer, as we are in a very shallow dive, but that need not worry us. This is the trim state that we want to be able to achieve at will whenever we need it. By killing the stall and setting the elevator we have effectively set the model to the right trim. However, straight and level refers to a range of trim states; Fig. 6.2 may help clarify the following:

Minimum Sink Trim (min. sink)
This is an S & L state that is just below the stalling position. Use this one for flying in lift and generally trying to stay up as long as possible. It is the slowest efficient speed that the model is capable of.

Best Lift/Drag Trim (best l/d)
This is just a little faster than the minimum sink trim. The model will come down quicker, but it will cover the greatest distance that it is capable of. Use this trim when flying from place to place in the sky, when returning from downwind, and when landing.

Penetration Trim
Faster than both of the above but not a dive, this trim is used to go places fast, such as when returning from downwind or escaping from a patch of sink.

These are the important trim states to learn and a competent

Fig. 6.2

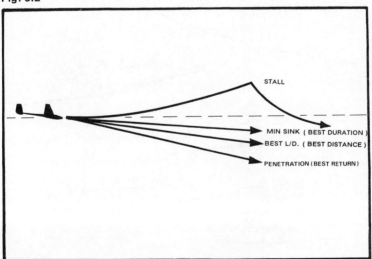

STALL

MIN SINK (BEST DURATION)

BEST L/D. (BEST DISTANCE)

PENETRATION (BEST RETURN)

flier will move from one to another according to the conditions. Each model will react differently, with smaller models having clearly identifiable trim states, though larger ones will show little difference in speed between the min. sink and best l/d conditions. A common mistake among beginners, and sometimes experts when they are nervous, is to keep pulling in up elevator to extend the flight time or try to make the model travel further. This should be resisted, as all it does is upset the trim of the aircraft and make it fly inefficiently. Leave the elevator alone when it is not needed for a manoeuvre.

TURNS

The first ones to learn are gentle turns onto new headings. Feed in some rudder in the desired direction. As the model turns you will see its nose drop, so feed in a little up elevator, which will keep the nose up and stop the wing on the inside of the turn from "digging in" and tightening the turn into a spiral dive. The tighter the turn is the greater the amount of up elevator required. Once the model is turning ease off the rudder; some gliders will need it held in to maintain the turn, some will need no rudder at all, most are somewhere between. As the model reaches its new heading apply opposite rudder and ease off the elevator as the model flattens out of the bank. Watch out for the model zooming up as it

Fig. 6.3

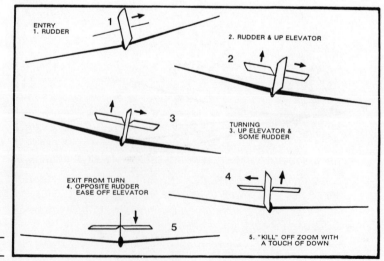

ENTRY
1. RUDDER

2. RUDDER & UP ELEVATOR

TURNING
3. UP ELEVATOR &
   SOME RUDDER

EXIT FROM TURN
4. OPPOSITE RUDDER
   EASE OFF ELEVATOR

5. "KILL" OFF ZOOM WITH
   A TOUCH OF DOWN

comes out of the turn and be ready to apply a dab of down to kill the zoom. The sequence is shown in Fig. 6.3:

ENTRY Rudder then Elevator (1 & 2) – TURN Ease off Rudder (3)
EXIT Apply opposite Rudder – Ease off Elevator (4) – Kill Zoom (5)

ZOOMING

Most manoeuvres will end with the model flying faster than the trimmed speed and failure to correct this often leads to a stalling situation. At the end of the manoeuvre watch the model as it starts to fly straight and if it starts to rise up on the excess speed wait till it reaches the correct speed and give a quick dab of down elevator to settle it at that airspeed. You are effectively preventing a stall and the process is the same.

THERMAL TURNS

Changing direction is relatively easy, but maintaining a continuous turn is more difficult. The procedure is the same as for a normal turn except that the elevator control must be more precise. Ideally the model should go round nicely without a great loss of height and without stalling in the turn. It takes practice. If the model stalls or wallows (goes mushy) in the turn ease off the elevator. If it stalls badly it may tip-stall.

ORIENTATION

There are two problems here: gauging the flight path at distance and when flying towards you during landings. At height it can be difficult to know if the model is flying towards you or away, and the best answer to this is never to take your eyes off the model. This will allow you the benefit of knowing what the last known heading of the model was and given this you should be able to work out its current heading.

Orientation on landing is a perennial problem with beginners, the real answer to which is practice. As the model comes towards you it will appear that the controls have become reversed, right stick makes the model turn to your left and vice versa. While you are learning try this method: when the model comes towards you move the stick towards the wing that you want to lift, i.e. if you want the model to go to your left move the stick towards the right wing so as to lift it. Likewise, if there is a danger of a wingtip digging in move the stick towards that wing to lift it. Only use this way of thinking whilst you are learning and only when the model is coming towards you. After a while you will unconsciously adapt to the problem and it will become easy to fly towards yourself. I actually found it hard to write this paragraph because I take this aspect of flying so much for granted. Give it time and it will come.

TIP STALLS

In a tip stall the model will rapidly drop a wing and dive. Apply opposite rudder to get the model out of the tight turn that it is in and recover from the dive.

THE SPIRAL DIVE

The result of not using enough elevator or tip-stalling and delaying recovery is a spiral dive. Effectively it is a steep turn and recovery should start with full opposite rudder and a normal recovery from the dive when the model is flying straight.

SPINS

If your model has its C.G. in a forward position it is very unlikely to spin. Thermal gliders do not spin easily, they prefer to spiral dive. The few occasions that I have seen one spin were after mid-air collisions. If it does happen put full down in to get some airspeed. When the model starts to go into a spiral dive recover in the way mentioned above.

HIGH SPEED FLIGHT

The problem here is that our controls are set for slow speed and the extra speed makes them very powerful. They will become very twitchy and rate switches are useful here. It is easy to get the model porpoising on the elevator, so treat it carefully. The rudder on a rudder/elevator model should be used progressively as energetic use may roll the model on its back.

OVERHEAD FLYING

Whenever possible, fly the model so that it does not go overhead. This means that your operating areas occupy a 360° sweep around you up to an angle of 60° to the horizontal, i.e. anywhere but above your head. Doing this allows you to keep the horizon in your field of vision, which means that you always have something to relate the flight path of the model to and you know if it is diving or stalling. As you fly above your head this information is not available to you, which causes worry and usually some flustered flying. The procedure to follow is to trust the model, since if it is trimmed correctly there is no reason why it should do something odd just because it is overhead. If you are thermalling get the model nicely set in its circling turn and simply hold the controls stationary as it drifts over. Do not panic, but if the model is obviously going wrong fly out of the overhead position to a position where you can get control of it.

ROLLS

Rolling a thermal soarer bigger than 2m span is not recommended practice. For a rudder/elevator model the trick can be achieved thus: dive into wind to gain speed, pull up and apply full rudder, then as the model goes inverted apply full down elevator to keep the nose up and release the down as the model comes upright. The result is usually untidy but fun, and if it goes wrong the model usually just screws out in the inverted phase.

LOOPS

This is the easiest aerobatic manoeuvre with a thermal soarer. Simply dive into wind to gain speed, pull in the up elevator and hold it there as the model goes round. Let the model climb on its excess speed and catch it in the usual way and not a lot of height will be lost. If it goes wrong the model will either corkscrew out the side of the loop or do a severe (hammerhead) stall, the probable cause being a touch of rudder at the wrong time or insufficient airspeed.

Having enjoyed the above you will now be down to a height where landing must be considered.

An open class soarer heads for the landing circle in a very steep approach using full brake.

7 LANDING

All good things must come to an end, and a flight must have a landing. When you are learning, wait until you have some confidence and control over the model before attempting to land. The secret of a good landing is a good approach and to get the approach right you need to be able to fly accurately. The better the approach the easier the landing.

LANDING APPROACH

As in full size practice, a model should be landed in a systematic fashion, at least until experience allows you to perform otherwise. The first thing to do is decide when it is time to start the landing approach. Watch the other fliers and see at what height they start to think about the problem, and get them to talk you through their flight. It is difficult to be specific because flying sites are so different, but start to work on the approach at about 75 to 100ft.

A standard square approach, Fig. 7.1, starts from an upwind position, with the model facing into wind. Trim the model for best l/d (i.e. put some down trim in to ensure that there is always sufficient airspeed for control) and turn so that it flies crosswind

Finger-type airbrakes are used in this *Proton* **to adjust its landing approach.**

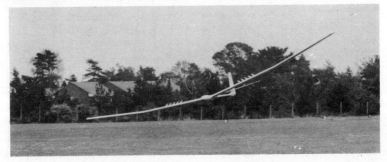

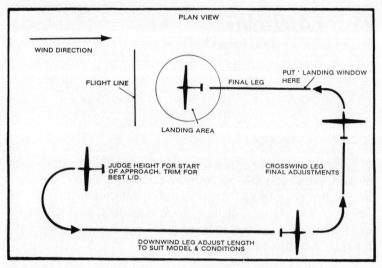

Fig. 7.1

for a short distance. Next turn downwind and fly to a position behind (downwind of) the flight line. Now fly crosswind until the model is directly behind the flight line and directly downwind of you. Turn the model into wind and point it towards you, and let it penetrate forward to land at your feet.

Fig. 7.2

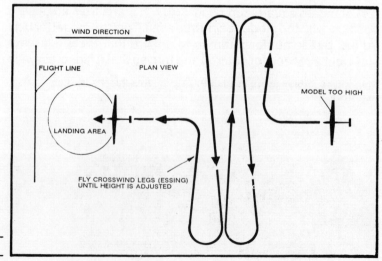

JUDGING THE DOWNWIND LEG

That sounds simple and it is until the windspeed and the model's height and glide angle are mentioned. The approach is adjusted on the downwind leg, the most important part to judge and the one that is difficult for experts on occasion. The downwind leg determines the height and length of the final leg (the run in to the landing). The windspeed plays a crucial part, as if it is calm the model can cover a lot of ground on the final leg, but if it is windy it will have to penetrate into the wind and will thus be limited in the distance that it can fly. Only experience with the model in a variety of conditions will tell you how the wind will affect it.

The next factor is the height of the model. Obviously if it is quite high you can go further back, but if it is quite low you must turn crosswind earlier. The glide angle of the model will determine the ground-covering ability of the model and this must be learnt by experience. It will improve with the addition of ballast.

UNDERSHOOTING

While you are learning you will land almost anywhere else but where you want to be, and this is perfectly normal. If you find yourself landing downwind of yourself (undershooting), do not worry, put the model down as best you can and enjoy the walk. Since this is a likely occurrence leave yourself plenty of space downwind to land in, never cramp yourself, leave your options open. If you are launching near an obstacle walk forward after launch to give yourself room; be aware of your surroundings.

If you have undershot analyse why it happened and adjust your technique to eliminate the problem.

OVERSHOOTING

This can be more serious if the model is far too high and likely to fly into something at the boundary of the field. If you find yourself coming in too high you need to lengthen the final leg, and the way to do this is to fly in an "S" pattern (essing, for want of a better word!). Turn one way and fly crosswind, then turn back and fly across wind again, turn back and fly crosswind, and so forth until enough height is burnt off and the final approach can be resumed. Fig. 7.2 may help. Do not let the speed drop off during this manoeuvre as the model will be close to the wind shear and you do not want it to drop out of the sky without warning.

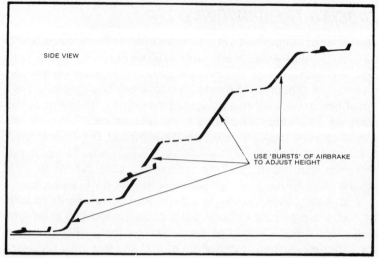

SIDE VIEW

USE 'BURSTS' OF AIRBRAKE TO ADJUST HEIGHT

Fig. 7.3

If the model is high for the approach and there is room to let it continue, land it in front of the flight line, again learn by the experience and adjust your technique. Landing has to be learnt the hard way.

AIRBRAKES

The use of airbrakes dramatically improves the accuracy of your landings. They can be used anywhere in the landing circuit to adjust your height but are of most use on the final leg. If you are too high, deploy the brakes to increase the sink rate. When you feel that enough height has been lost drop them back down and continue the approach. This sort of adjustment can be done several times on the final leg. Watch out for the trim change when the brakes are opened, for the model will usually pitch downwards and need some up elevator held in to prevent it from diving. Get used to the brakes at height so that you feel confident about their action. After a while you will habitually come in a little high and use the brakes if necessary. See Fig. 7.3.

THE LANDING WINDOW

A concept that I have found helpful is the use of an imaginary window in the sky. This delusion is placed at the correct height and distance downwind so that if the model flies through it at the right speed it will land at my feet. Only experience will tell you its

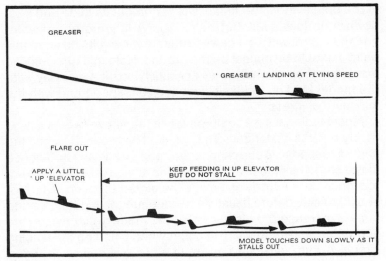

GREASER

'GREASER ' LANDING AT FLYING SPEED

FLARE OUT

APPLY A LITTLE
' UP 'ELEVATOR

KEEP FEEDING IN UP ELEVATOR
BUT DO NOT STALL

MODEL TOUCHES DOWN SLOWLY AS IT
STALLS OUT

Fig. 7.4

location, but once found it will tell you early if you are too high or
low.

LANDING

If left to its own devices the model will fly straight and contact the
ground at flying speed, which is commonly called a "greaser"
since the model skids along the ground for some distance. This
style of landing is probably the safest for a beginner, since there
is little likelihood of a stall close to the ground and positive
control is maintained throughout the final few feet. Fig. 7.4
illustrates.

The better way is to flare out. Switch out the airbrakes. Let the
model fly to within a couple of feet of the ground and gently ease

**A typical spoiler
installation. This
one uses magnetic
hold-down, over-
come when the
servo is energised.**

in the up elevator. The model will flatten out its glide and slow down. As it slows it will start to sink. Apply a little more up elevator and the model will slow down further and raise its nose. At this point it should be almost on the ground. Hold this trim until the model touches down. This is very satisfying on a nice day, with the model landing at its slowest possible speed and with the minimum of stress.

Most landings are a compromise of the above two. In a wind flaring out has its dangers. The speed at flare-out is low and the control response is correspondingly poor, while any turbulence may disturb the model and crash it. The obvious mistake is to put too much up in initially and make the model zoom up. If it does, stop it quickly before it stalls and continue the landing. Flaring out will increase the distance that the model will travel on the approach, so take this into account when planning the landing circuit. A long flare-out can be used to overcome an under-shooting situation.

TURNS CLOSE TO THE GROUND

Obviously a model with a large span must be turned much more cautiously close to the ground to save it from digging in a wing tip. A less well recognised danger is the wind shear (Chapter 10), which can make the into wind turn nasty. The wing that drops for the turn may go into a slower moving air mass and as a result it can stall. The upshot is that the model drops down sideways onto the stalled wing. Avoid this by doing shallow turns with big models close to the ground. Little models are far less likely to be affected.

ALTERNATIVE APPROACHES

It is unlikely that you will always be upwind to start the approach. If you are crosswind then you will have to judge the height and distance for the start of the final leg from a crosswind approach, and "essing" may well help. An approach from downwind will need some judgement of the actual start of the final leg, but "essing" will again help to adjust the height. Alternatively fly to a crosswind position and start a crosswind leg from there.

As experience is gained the approach will become more loose and less deliberate, but it will still be based upon the principles learnt from a standard square approach.

OBSTACLE AVOIDANCE

One problem with model flying is an effect like tunnel vision. You can only see your model and no others exist. This is acceptable in clear air, but in the landing circuit you should be aware of the positions of other models so that you can avoid them and not be guilty of a mid-air. If you look like being hit by someone who has not seen you, dive, zoom or turn out of the way and argue the toss later.

Ground obstacles should be checked for early in the approach and during the final leg. Beware of trees and bushes because your depth perception is not very accurate. If someone wanders into your path shout a warning: "heads" is usually sufficient with fliers, but something a little longer and more direct may be called for with the general public. If this fails use the extra airspeed that you have by virtue of trimming in down at the start of the approach. Zoom up and over the obstacle and recover as quickly as possible before the stall. Turning may be possible but often leads to a sudden arrival. Do not be severe with the subsequent site safety lecture if a member of the public is involved – it might jeopardise the site.

Good accurate landings take a long while to develop and while familiarity with the model helps a lot, expert guidance is required. Don't ever be tempted to fly by yourself before your instructor thinks that you are ready. Once the flying aspects are being mastered he will probably allow you to get onto the approach and suddenly just let you land. It's a magic moment. Once that is mastered the launch can be tackled. My instructor just handed me the transmitter, said "Hold this a moment", grabbed the model, said "Are you ready?" and launched. Before I could protest the model was at 100ft. and going up straight as a die. It was a good job that I did not have a heart condition. Learning to fly is possibly the best phase in modelling, so enjoy it for its challenges and personal triumphs.

D

8 THERMALLING

Once you have got past the initial learning stage and are regularly flying solo, things can get a little dull. Now is the time to move to one of the most fascinating and tantalising areas of soaring, thermalling. Without doubt it is the element that keeps people flying for years without loss of interest. The challenge is enormous, as no two days' weather are ever the same (certainly not in Britain) and lift can be totally unpredictable or as regular as clockwork. The same flying field will work in vastly different ways with changes of wind direction, temperature and weather. Thermalling is a subtle game that makes you work with the forces of nature.

EQUIPMENT

Now is the time to build that second model. The old trainer may well still be usable but its performance may well be insufficient for good thermalling flights on days with weak lift. Something in the region of 100 to 124 ins. (2.5 to 3.5m) is now required, of fairly robust construction and easy to repair.

The other piece of equipment that you need is a stopwatch of some sort, for which the reasons will become clear later. A hand-held stopwatch is fine but it must have some definite means of indicating that it has been switched, such as a bleeper or switches that click when operated, since you do not want to have to take your eyes off the model to check that the watch is running. In addition it should not be liable to switching itself when it is bouncing around as you move around with it hung around your neck. A cheaper alternative is to use a wristwatch with a stopwatch function, but don't leave it on your wrist, the last thing that you want to do is to be fiddling with tiny buttons just after the model has left the line. Take the wristband off and stick the watch onto a convenient area of the transmitter or the eye shield (use double sided tape). A further possibility for those of you who feel like experimenting is to permanently instal a watch module in the transmitter case and fit buttons on the outside of the case. Care, good eyesight and a fine soldering iron are required, and it would pay you to consult with the manufacturer or his agent about whether such a modification affects the warranty on a new transmitter.

PRELIMINARY WORK

You must firstly make sure that you have a properly trimmed model. The Centre of Gravity should be in the 30 to 40% range and the model should be capable of being flown straight and level "hands off" at most speeds. The next step is to really get to know the model, its little quirks, its stalling behaviour, its low speed handling and its circling abilities. This last ability is essential. If the model doesn't circle well fiddle around with the C. of G., the dihedral or the fin and rudder size until it does. It should be stable in the circling mode and should be capable of being held there either "hands off" or by the use of elevator and rudder trim. The degree of bank angle required for efficient circling varies from model to model and while some need to be circled fast at about 30° of bank, others will float around slowly at around 10°. Generally the heavier the model the faster and steeper it must be flown in the turn. The more work that you do to get the model to fly right, the easier the task of lift spotting becomes.

During this familiarisation and trimming period you should get into a set routine with the stopwatch so that a reasonably accurate timing is taken of each flight. This will give you a rough idea of what constitutes a "standard" flight time for the model. This piece of information is important since you can now judge the quality of your flight according to its duration. If you get a low time try to ascertain why; was the trim O.K.? Did you overcontrol the model? Was there sink about? Are the conditions not suited to the glider? Was it over or under ballasted? And so forth. If you get a good time were you lucky? Could you have done better? Did you actively seek the lift? and so forth. The use of a stopwatch allows you to study the variables and improve your technique, and the use of a well trimmed and familiar model reduces the number of variables further so that one is left with the atmospheric conditions to deal with. There is an additional advantage to using a stopwatch; it allows you to keep a record of your best flight time and you then have the constant spur of trying to beat your previous record.

READING THE SIGNS

Three definitions:

Lift – rising air; can be propagated by a variety of causes such as hills, fires, factories, thermals, mixing air masses and so forth.

Thermal – a mass of warmed air that is rising up through the atmosphere due to its lower density.

Sink – cold air that by virtue of its higher density is falling out of the atmosphere.

It must be understood that not all flat field soaring is achieved using thermal lift. You use anything that you can get hold of – little bits of slope lift from roofs or ridges, scraps of warm air off parked cars, rising air caused by a cold front shovelling under a warm front, etc., as well as the classic thermal. Thermals of the "cumulus producing type" tend to be fair weather creatures at the sort of heights that we habitually fly at. Higher wind speeds tend to shred the lift into streaks before it can form up properly or pile it into large patches and the characteristics of the lift then become complex and our problems are compounded by being tied to one spot on the ground and having to fly upwind all the time.

What to look for? Accurate observation and practice are required and the accuracy of your observations will depend largely on your knowledge of the model and flight time analysis.

This flier is reading the signs to decide when to launch. Model is a *Clean Machine* **wing on a** *Proton* **fuselage.**

Fortunately the signs for strong lift are fairly obvious and those that affect the model can be broken down to the following:

1. Apparent loss of control

If the radio and model are O.K. and the model seems to be bucking about as if it is out of control then it is very likely that you have encountered the turbulence to be found in a strong thermal. Most beginners mistake such turbulence for radio problems and immediately fly out of the area, whereas the correct course of action is to stay with it and search out the best area of lift to exploit.

2. An unexpected course change

If a properly trimmed model veers off to one side for no apparent reason then it has touched the edge of a thermal. If you imagine a thermal as being round, as in Fig. 8.1, then what has happened can be seen from part 1. One half of the wing flies into the rising air in the thermal and is lifted, which yaws the plane away from the lift. It pushes you out of the lift and points you away from it. As a general rule if the model is reluctant to fly into a particular piece of sky then that is where you want to be. There are two possible courses of action, and the fastest one is shown in part 2 – you immediately turn against the unexpected course change and fly for a short distance at 90° to your original course, and then start to circle back into the area where the thermal is thought to be. Try

Fig. 8.1

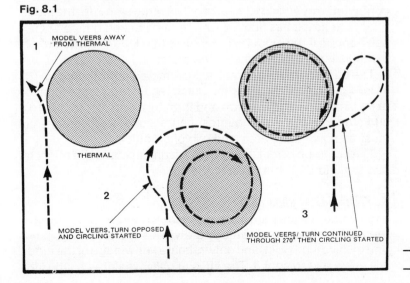

a couple of circles and if the model continues to descend at the normal rate, fly away, it was probably a gust. If on the other hand the model rises, stay with it and ride it out to a suitable height. If the model does not rise or for that matter stops sinking still stay with it. No height is being lost and the thermal may well be building up strength, but even if it isn't then nothing has been lost and you can move off elsewhere.

The other way of entering a thermal in this case is to let the turn continue for 270°, as in part 3, then flying at 90° to your initial course, enter the back of the thermal area and start to circle. Which way you choose depends upon the model and the strength of the thermal. If the rudder authority (power) provides a fast response then the first method is to be recommended and it is the most common one. If the model is slow in response then the second method may actually be quicker and lose less height since the model stays in the same bank all the time and does not have to change direction.

The common mistake made when the model is veered off course is to correct the course alteration and fly straight on, or to continue in the new direction.

3. Vertical disturbances

If you fly right through the centre of a thermal the model may do one of two things. If trimmed close to the stall it may actually stall for no apparent reason. On two occasions over roofs I have actually had a lightweight model do an involuntary loop when it contacted very strong lift head-on. The other sign is for the model to appear as if it has flown over a "hump backed bridge", which can happen if the model is trimmed to fly a little more quickly, or the lift is light. Fig. 8.2 shows the effect.

The best course of action is to turn immediately through 270°, as in Fig. 8.2, and fly across the "bump" again to give confirmation of its existence and location and then turn into it from the front and start circling. Such "bumps" can occur at low altitudes but must be flown accurately and exploited quickly.

The common error is to fly through and mistake the effect for bad trimming, i.e. too close to the stall.

4. Flattening of glide path

If the model's glide path flattens out then there is sure to be lift in the area, so try circling immediately. This is the most common and elusive sign of lift, and only intimate knowledge of the model will allow you to spot it. The lift is usually weak but a little lift is

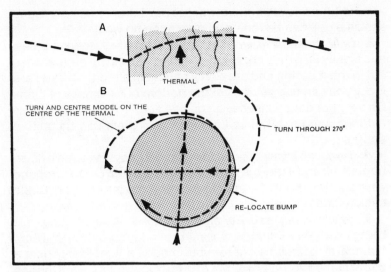

Fig. 8.2

better than nothing and it does allow you to hang around longer and wait for better conditions.

5. Improvement in model penetration

A model will frequently speed up when it flies into "good air" and this is particularly marked on windy days when you may be struggling to stay upwind. You can either take the penetration or ease back on the elevator and take the extra height.

One of the nicest things about soaring is that it makes you more aware of your immediate surroundings. The following environmental and weather clues are worth watching for, the list is only partial.

ENVIRONMENTAL CLUES

1. Other Models You can apply all of the above to the flight pattern of other models and benefit from the experiences.

2. Birds Gulls and birds of prey are probably the best lift indicators, but crows will thermal if the lift is strong. Swallows can be deceptive since they feed on airborne insects. Insects do rise with thermals but they also fall with the sink. I once saw a gull trying to join a thermal late, having been attracted to it by a flock of his mates who were going up fast. The swallows were belting around at low level, and he went toward them and only found sink. If a gull can be fooled what chance do we have?

3. *Light materials rising* Thistledown, mayflies, soap bubbles, etc. signify lift if they are rising. Mayflies rise as the ground dries out in the morning and are easily caught in a thermal. If you are lucky you can find yourself at the bottom of a chimney of them. On very hot days "dust devils" can be seen to pick up dry grass cuttings and dust, but soaring in one of them can be quite a struggle.

4. *Towman's signal* A good towman will know if you have flown through lift and if there is a set of signals worked out beforehand your spotter should be able to give you the information. Circle back to find the good air.

5. *Unexpectedly easier or higher launches* When a bungee is being used you will find that for any given day's weather there is a "standard" launch height for your model. If it suddenly gets a better launch you may well have flown through lift. Get off the line and circle back. Similarly, if the towman has a much easier time or is even running back towards the model it is likely that he is towing through lift.

WEATHER SIGNS

1. *Clouds* At the sort of heights that we fly at clouds are only an indicator of the general conditions and the lift pattern. Sometimes you can use them to pinpoint thermal activity but they are not usually a reliable sign. The thing to look for is activity within the cloud itself, swirling, doughnut shapes, dark clouds with domed bases, holes in large cloud masses, signs of general movement. Lack of activity usually indicates neutral or sinky conditions. A common mistake is to look upwind for cloud signs when the cloud that may be affecting conditions at our levels is usually downwind of our ground position. The clouds upwind will be over our position in 3 to 10 minutes depending upon the wind strength, by which time the lift pattern in them may well have changed. A further problem is that activity at cloudbase may well not be having any effect at lower levels. Clouds give general information most of the time.

2. *Temperature increase or decrease* Since thermals consist of masses of air warmer than the surrounding air it follows that if you feel that the temperature has risen rapidly then a thermal is passing over your ground position. It may well pay to launch immediately and search for the lift downwind of your position, providing that your model can penetrate back in the prevailing wind conditions. Conversely, a rapid drop in temperature indicates the presence of sink, colder air falling due to its greater mass. Do not launch in sink, wait for the lift that may well be following it.

3. Falling or increasing windspeed A decrease in windspeed is indicative of the presence of a thermal, and often this is accompanied by a temperature increase, but whether you launch or not is up to you. An increase in windspeed, frequently accompanied by a temperature drop, is a sure sign of sink.

4. Change in wind direction This is another sign of thermal activity and a good reason for having a ribbon attached to your aerial. The theory is that if the wind direction moves then the change in direction is pointing toward the thermal, i.e. if the wind veers to your right then the thermal is somewhere to your right. It does seem to work but as with all atmospheric signs it can be confused by other causes such as the arrival of a front.

The signs for light lift are much the same as for strong lift but much more subtle, and the ability to recognise and use light lift is the difference between a good and an average flier. However, it is essential that you get experience of strong lift before you can start to learn about light lift. Anything will go up on a really good day. I personally feel that our generally appalling conditions eventually produce better pilots.

MAXIMISING THE THERMAL

There is no point in locating lift if all you do is then meander aimlessly around; for the best results you must find the strongest area of rising air and stay in it. A thermal will try to stop you from circling in it by trying to flatten out your angle of bank, and the stronger the lift the greater the effect. So it follows that if you are circling and the model does not circle evenly then you are not centred on the strongest area of the lift. If the model flattens out you are near the core of the thermal, if the model tightens up then you are to one side of it, as in Fig. 8.3, assuming of course that you are holding the controls steady. The idea is to move the centre of our thermal circle towards the spot where the model is being flattened out most and try to centre on the core of the thermal. Once centred the model should be capable of being trimmed out for circling flight and you should be able to leave it to its own devices.

This method is fine for the "classic thermal", and is applicable

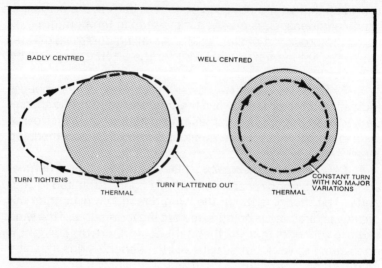

Fig. 8.3

to any patch of lift, although "broken" lift generally requires more care.

Lift close to the ground tends to be in fairly tight areas and will need to be exploited at a fairly high angle of bank, though generally try to keep the angle of bank to a minimum consistent with efficiency.

Windy days will often present you with scraps of lift. Since it may not be possible to thermal-circle due to the problem of getting back upwind, a different approach is required. Since the model is probably flying with a fair amount of down trim it is flying well above the stall speed and thus when you contact lift you can put some up elevator in, let the model rise and anticipate the stall by easing off the elevator. The model should gain some height whilst maintaining its position. The method is sometimes called "pole bending".

COMING DOWN

Having gone to all the trouble of finding lift we are now faced with a further problem, how to stop the model getting too high and how to stop it from being sucked out of sight. How high you fly will depend upon how well you can see the model, so as soon as you feel uncomfortable think about getting the model out of the lift. With well-defined thermals just flying forward out of the lift may be enough to put you into the sink next to the thermal and allow you to adjust your height back to an acceptable level before you go in search of more lift.

With large, powerful areas of lift it may not be possible to just fly out of the rising air and height must be shed by means of deliberately inefficient flying or loss of airframe efficiency. This is where a good set of airbrakes or spoilers that can limit the terminal velocity of the model are essential. I would recommend the installation of airbrakes of some sort on any model that has an efficient soaring performance, not just for pilot convenience but as a safety factor. It's not a good idea to be below a model that has fallen apart in the air due to being overstressed in the dive. With airbrakes or spoilers the procedure is to deploy them and try flying forward, when in moderate lift the model should start to sink at a reasonable rate. If this does not work feed some down in and try a braked shallow dive. If the lift is so strong that you are still not losing height then try a spiral dive, with brakes out to limit the speed. Without brakes you are very dependent upon the strength of the airframe. With a lightweight model probably the best method is a sort of hard banked circling that is almost a spiral dive, but with a strong heavyweight model a shallow dive may be attempted, or a spiral dive. Generally, the smaller the model is the tougher it is, and the more punishment it will take. The *Gentle Lady*, for instance, has been seen to dive steeply for long periods without any ill effects. Larger models with long thin wings (high aspect ratio) can easily be prone to flutter at high speeds and flutter is a model killer. Try to avoid a prolonged dive if at all possible.

The best method is to plan your ascent so that you do not put yourself in a position where you have to do anything in a hurry. Do not get carried away with the joys of the thermal hunt, stay calm, think ahead, fly to survive. Tomorrow is another day and you don't want to spend it in the repair shop.

There are no shortcuts to being able to thermal-soar well: experience, practice, observation and a well-tried model are all required. If you are keen and active you may well pick up the rudiments in a year, although more likely it will take a couple of years before you feel that you are moderately competent, and no one can claim to know everything about the subject. But the sense of satisfaction when you do find lift is tremendous. The thrill of the hunt, the achievement of new personal records – it really gets under your skin.

9 SLOPE SOARING

by Keith Thomas

If you are keen to start slope soaring and have turned straight to this chapter, a word of advice. Almost everything that Dave Jones has to say elsewhere in this book about thermal soaring applies in equal measure to slope soaring, so please read the whole book before getting stuck in! In "my" part of this book, I discuss only those areas of slope work where there are important differences from thermal practice.

In fact, if thermal soaring is what you really want to do, but you happen to have a nice slope close to home, then you can follow what Dave Jones says all along the line, and only read what I have to say on launching and landing.

MODELS FOR THE BEGINNER

If the starting point for the budding thermal flier is more or less standard, the same certainly does not apply to the slope. As slope lift varies from nil to extremely powerful, depending on the site and the weather, slope models cover an equally broad spectrum. On a balmy summer evening you will find the slopes littered with lightweight polyhedral "floaters" which would otherwise be confined to flat field sites. Push the breeze a little up the scale and you find a wider spread: floaters, three-channel semi-scale jobs, small-span two-channel general-purpose models, scale sailplanes and maybe a smattering of flying wings. Tighten the screws further, and the specialist pylon racers and small aerobatic craft are dusted off and brought out. Many modellers even keep specialised heavy weather gliders in the attic for the real blowy days, when heavily ballasted racers, deltas and flying wings come into their own. This vast range of types constitutes one of the great attractions of slope flying, but it does make the beginner's choice much more difficult.

If you are starting out in modelling, or RC modelling, in the spring or summer, your best bet is probably a general-purpose

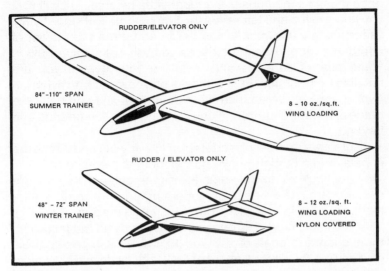

Fig. 9.1 and 9.2

thermal soarer of moderate size. A rudder/elevator model around 8ft. span would fit the bill (Fig. 9.1), preferably with a built-up wing and a wing loading around 8–10ozs. per sq. ft. There are dozens of such models available, both in kit and plan form. In general terms they are docile to fly, stay up well in light lift, and will give you many hours of basic practice provided that you don't attempt to fly them in strong winds. In typical slope conditions they are not particularly easy to land, so my recommendation is tempered slightly if your slope site has a restricted landing area. Airbrakes or spoilers solve the landing problem, but don't be tempted to try them at first – they just give you an extra headache. If you go down this route, only fly on pleasant days with light winds. On the breezier days you can sit at home and build your second model.

Rare are the winter days when we can fly our floaters. If your initial learning stage coincides with the bluster of autumn and winter, you need a different kind of glider – smaller, more robust, but similarly easy to fly, (Fig. 9.2). The standard choice used to be the Veron *Impala*, and if somebody offers you one of these excellent models ready built, grab it without delay. It is a rudder/elevator model, small at 54in. span, but it offers a capacious fuselage and can take a lot of harsh treatment. Personally I rate such small models as relatively difficult to fly, as they react quite sharply to the controls and are not easy to fly on a smooth course. Nevertheless, if you can keep it in one piece – which is not difficult as the *Impala* is a tough little machine – it will give you hours of fun and teach you the fundamentals of slope flying. Unfortunately the *Impala* was designed before the advent of affordable proportional radios, so you will need some help in fitting control surfaces and installing your gear if you are building from scratch. Built with care and covered with nylon, an *Impala* can survive an incredible pounding, and can handle a range of wind speeds from "floater" weather up to moderate and beyond.

There are a number of models of similar lay-out and size to the *Impala*, and although I have no personal experience of them, they are likely to do the same job as the original. As they are designed from the outset for modern proportional radio equipment, you should have less trouble installing the gear, (Fig. 9.2).

At the initial stage I strongly urge you to avoid the following: three-channel models of any type (Figs. 9.3 & 9.4) (two controls give you plenty to worry about at first), any model which is nice-looking (you will never be able to relax and concentrate on flying it for fear of marring the coachlines), anything unusual (canards, deltas, flying wings, biplanes), anything expensive (you are bound to break it) and anything you cannot repair. I feel quite strongly that a foam wing is a rotten bet for a beginner. As Dave Jones has pointed out, a standard foam wing has too little strength at the root and too much weight at the tips. Heavy wingtips have an alarming effect on a glider's control response,

Fig. 9.3 and 9.4

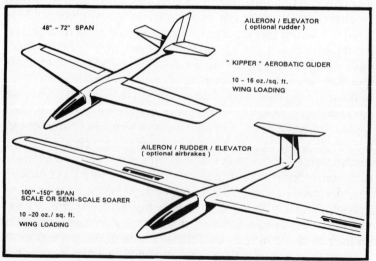

48" - 72" SPAN

AILERON / ELEVATOR
(optional rudder)

" KIPPER " AEROBATIC GLIDER

10 - 16 oz./sq. ft.
WING LOADING

AILERON / RUDDER / ELEVATOR
(optional airbrakes)

100" -150" SPAN
SCALE OR SEMI-SCALE SOARER

10 -20 oz./ sq. ft.
WING LOADING

Mike Freeman and his hand-launched flying wing *Zippy* **slope soarer, featured in** *Radio Modeller*.

making it difficult to turn, and equally difficult to stop turning. A properly constructed built-up wing, on the other hand, ensures delightfully prompt rudder response. The model will turn at once, and come back to straight-and-level nice and quickly when you need it. The other drawback to a foam wing is the problem of repairing it.

Make no mistake about this: you will break your first trainer, probably regularly, and you will have to mend it. Repairing damage to a foam wing is tricky, as the foam is such an intractable material. Your wing ends up heavier, thanks to the glue, glass cloth or bandage, and you have to add more unwanted weight on the other panel to balance the repair. Your unwieldy model becomes yet unwieldier and you crash again. Guess where it will break . . . yup, just next to the repair. In any case, if you build a ribs-and-spars wing you will know just how it went together, and how best to repair it.

SLOPE SITES

Two problems here: the physical characteristics of the slope, and what we might term the political implications.

It is important to find a reasonably effective slope and wait for a "good" day for your first attempts. Most people find flying a slope soarer difficult at first – some never get the hang of it – and it is far from helpful to try out a new model on a day with insufficient wind at a slope which is not "working". What do I mean by "good", "effective", "insufficient" and "working"? All these terms are hard to nail down, and only begin to make sense once you have a modicum of experience. For this reason it really is highly advisable to find an experienced slope flier to help you right at the start. As a beginner you have no criteria by which to judge a site or the prevailing conditions, and your chances of success are very small. Risk it by all means if there is really no option – I'll give you some hints later on – but you have to accept that you are courting disaster. If you are not sure of the site or the conditions, you should wait until you see a seasoned slope flier actually launch and fly and land successfully, preferably flying a model similar to your own, before having a go yourself.

But you may have bought this book precisely because the advice and experience of others is not available to you, so don't let me put you off. The perfect slope site looks like this: you approach it by means of a road to the top which goes nowhere else. The top of the hill will be perfectly flat, covered in thick, rather long grass, and devoid of trees. The slope face will be steep – preferably 45° or so – smooth, and similarly featureless. It will dive down many hundreds of feet before curving out in a south-westerly direction into an infinitely long plain. The top of the slope will be nicely rounded, but you will find a depression just behind the lip where the effect of the wind is mollified and your tea runs from flask to cup without being blown away downwind. On the "political" front, the site will belong to nobody, but somebody will have erected an official sign stating "danger area – no hang-gliding – suitable for slope-soaring model gliders only".

You may be lucky on your search, but it is more likely that you will end up with something like this – my local site until I moved house recently – a 10 acre field, flattish at the top and sloping at about 1 in 10 down to the valley. Height from lip to floor about 100'. Telephone poles stalking diagonally across the landing area, itself bounded at the rear by a tall fence surrounding a derelict industrial site. Trees all round the field. Slope facing due west, gentle bowl shape. In short, everything wrong. Was there anything in its favour? Indeed there was – it was 100yds. from my back door! The lift was patchy, totally unreliable and weak at

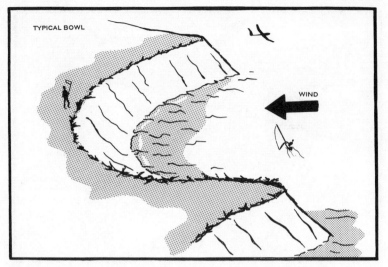

Fig. 9.5

best, but I had many hours of enjoyable and pleasantly difficult flying there.

The basic requirements are a steep slope, somewhere to land and permission to fly. Everything else is a luxury. However, the shape of the hill is important. There are three basic types of hillside which are of use to us as slope sites: the bowl, the straight ridge, and the pimple.

The term bowl (Fig. 9.5) is loosely applied to any hillside which is concave when viewed from above. It might be essentially a straight ridge with a gentle curve at either end, or a deep valley with long, projecting arms. Either will work well when the wind is exactly central, i.e. square on to the centre of the slope, with the deep bowl producing tremendous lift from the funnelling effect of the valley. On the other hand the shallow bowl will tolerate slightly off-centre breezes, whereas the deep bowl will not.

The straight ridge (Fig. 9.6) is a rare beast, but the best ones offer superb gliding when the wind is "right". The pimple (Fig. 9.7) is also quite rare, and quite difficult to fly, but if it is the only site available don't pass it up. I have had many hours of good flying from a particular hill of this type, even though the top is scarcely broad enough to land a model on. At this site most faces of the hill are slightly dished, and generate lift over a much broader swathe of air than the length of the slope itself.

An aspect of any slope site which is more important than it may at first appear is the transition from slope to plateau. If the

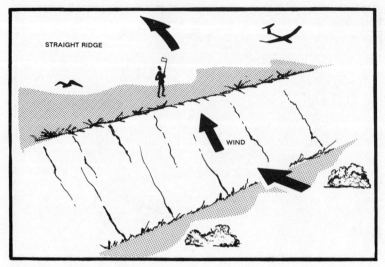

Fig. 9.6

slope ends in an abrupt ridge, you may find turbulence very disturbing in certain conditions. It is not possible to predict these effects, but as the bumpy air occurs just where you want to land your model, it is undesirable to say the least. A smoothly curving transition is a far better bet.

One of the great challenges of slope soaring is wringing the best out of a third-rate site. You may well learn to fly at a good slope a thirty mile drive distant, but I bet you can find something flyable closer to home once you have a little experience. You may not appreciate this at first, but it is just as exciting to keep a model successfully aloft in tricky conditions as it is to exploit massive lift on a good day at an efficient site.

CONDITIONS

Slope gliders can be built to handle anything the elements can hurl at us, but your first model will only be able to cope with quite a narrow slice of the spectrum. Assuming that you have built a model of reasonably small size – 5–8ft wingspan – and that it has a wing loading between 8 and 14ozs. per sq. ft., you will want to wait for a day with a moderate breeze. I bet you haven't got an anemometer strapped to your hat, so here's what I think of as "moderate": standing on the edge of the slope, gazing out straight into wind, you feel a constant, cooling flow of air straight in your face. It might make your eyes water very slightly, but it will

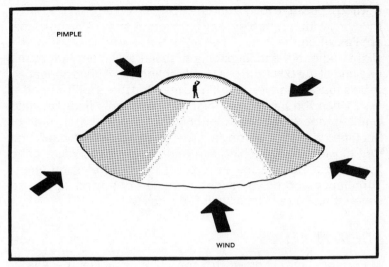

Fig. 9.7

not blow your hat off. If the airflow is patchy, the direction of the breeze liable to vary, then don't fly. Your model will surely go to the bottom of the hill. If you have to screw your eyes up and hang on grimly to your hat, then don't fly. Your model will equally surely disappear behind you.

Okay, you have a steady, moderate breeze and an inviting slope. What about the wind direction? The experienced sloper will enjoy "taking on" a slope when the wind is not square onto it, but at the start you must not risk it. The wind should blow more or less exactly at 90° to the slope face. In most cases it is quite easy to check this, using your ears. Stand facing the wind, remove any headgear, and rotate your head slowly. As your downwind ear moves into turbulent air you will hear a loud rushing noise. Turn your head back again until the same thing happens on the other side. Half-way between these two points and you are staring the wind right in the eye. A wisp of knitting wool tied to the end of your transmitter aerial does a similar job, but nowadays most frequency pennants are too heavy to do anything other than whizz round and round.

Establishing the wind direction presents no problem if the slope is a straight ridge, or part of a "pimple". The main reason I mention it here is because of the problems involving wind direction at bowl-shaped slopes. Some of the best slope flying sites are bowl-shaped, because of the intensifying or funnel effect they create. With a dead-centre wind on a deep bowl, you will find it possible to fly from one tip to the other, in strong lift all the way, with no problems of turbulence at all. Alas, this only applies when the wind is just right. If the wind is blowing at a slight angle to the ideal direction, the air will often flow round one end of the bowl and change direction into it, thus appearing to be square-on as you face it. You may find that the lift is good at first. Once the model gets high, you suddenly find that "into wind" is not where you thought it was, and the lift dies or becomes bumpy. If the deviation is small, you will probably find that moving the launching site to one side gives you a good flying area. Beware the far side of the bowl at such times! Turbulence and downdraught in the lee of a wind can be so severe that you will think your radio has failed.

GOING IT ALONE

Many modellers are drawn to slope soaring precisely because it can be and often has to be a solitary activity. If you plug away at anything on your own you run the risk of boredom setting in, unless the activity offers a really wide scope. Slope soaring is not uniquely good in this respect, but it is hard to beat. If you multiply the different categories of model by the number of different tasks you can set yourself, you end up with a vast number of permutations. Provided that you don't get stuck in a rut, the solitary slope modeller will never run out of ideas.

Slope flying, then, attracts the non-gregarious, and for this reason the standard advice offered to every beginner – to join a club – is not always the answer. Even if you are the type who would willingly soak up the advice and experience of others, you may live out in the wilderness where clubs are available to sheep only. People can and do learn to fly without outside help, but you must expect it to be very difficult, model-consuming, and dispiriting. You are effectively emulating the Wright brothers: you have an untried aircraft, no skill, untrained reflexes, and probably a flying site of unknown quality. This mixture is a recipe for disaster, and before I offer any bits of advice, I urge you one last time to seek help. Not to join a (possibly non-existent) club, but to travel with your new model and equipment to a known site, however distant, and seek help there just for a day. In the last resort, consult the Yellow Pages, find your nearest model shop, telephone the manager and ask where the local slope-soaring group operates. Go to the site and ask somebody who looks skilful with the sticks to check over your model and your gear, and to test-fly it for you. Most people are flattered by such an approach, and I have met only a handful of modellers who are selfish enough not to want to share their knowledge with a beginner. If you get a chance to fly the model yourself, perhaps with the aid of a buddy-box, so much the better. Watch your instructor like a hawk; you could even take notes or keep a surreptitious tape recorder running while you grill him. You can learn an immense amount of priceless lore from such a meeting.

From now I assume that you have no outside assistance at all. And the best of luck to you.

PREPARATIONS

You have built your model, following the instructions as precisely as possible, and have installed the radio receiving components correctly. You have checked the wings, tail surfaces and fuselage for warps. You have balanced the model at the point shown (a little further forward is usually permissible, further aft is not). You have checked that the wings are balanced, i.e. the model is balanced about its centreline. The model's component parts are adequately lashed together. The model is within the weight limits stated by the designer. When you move the rudder stick to the right, the rudder moves to the right. When

Keith Thomas himself with his winning quarter-scale model at a White Sheet scale meeting

you pull the elevator stick back towards you, the elevator (or rear edge of an all-moving tailplane) rises. At full stick movement, neither control surfaces binds or jams up, and the servos remain quiet. A loud buzzing noise indicates a mechanical obstruction which must be eliminated before you fly. Your batteries are fully charged (nickel-cadmium packs) or freshly purchased from a large store (with a high battery turn-over). You have travelled to the site in a car with enough petrol to get you home again, your flask is full of tea. The wind is blowing straight into your face. Your knees are banging together.

LAUNCHING

Hold on a moment. You don't know what you are going to do when the model is up in the air, so don't launch yet.

GIVING YOURSELF A CHANCE

That's more like it. If your site has a really large, flat landing area, then I recommend that you try a hand-glide before committing the model to the sky. Walk downwind at least one hundred yards, switch the transmitter and receiver on, and check that the trims and control surfaces are central. Are the wings and tail square on the fuselage? Grasp the model under the wings with your right hand (left if you are left-handed), and pick up the transmitter in your left. Holding the model just above your head, with the wings level and the nose inclined slightly downward, trot a few paces into the wind and push the model forward forcefully, as if throwing a heavy knife into the ground ten yards in front of you. Get your right hand back onto the transmitter without delay, and watch what the model does. If you have been careful all along the line, and if the model designer knows his onions, your model will glide in a reasonably straight line to the ground. If it shows every sign of doing this, don't touch the sticks! It may wander to one side, it may land a little heavily, it might glide more steeply than you wish, but if it looks safe, leave it alone.

If, on the other hand, the model looks in trouble, you have to try to sort it out. The most likely problem is a dive or a stall. If it rears up sharply (a stall), you have to push the elevator stick forward – away from you – in an effort to prevent the model coming to a halt in the air and crashing. As a complete beginner you have my congratulations if you managed that feat, as quick

Moment of concentration (and Body English!) from the flier as a helper launches his model.

reflexes are everything in this situation. It is more likely that it will hit the deck before you remember the right response. With luck, the damage will be slight or non-existent.

If the model dives straight into the ground, the correct response is to pull the elevator stick back to pull the nose up, but in this case the model will undoubtedly hit the ground before you realize what is happening. Again, you will be unlucky to sustain serious damage.

The model may turn to one side strongly enough to touch one wingtip on the ground and swing round. This will not usually cause damage.

In any case you must find out the cause of the problem. There is not much that can be wrong. With a stall or dive either the balance point is wrong (check on the plan – always balance a model indoors to avoid the wind messing it up), the wing and/or tail incidence is wrong, or the elevator was not neutral. In terms of stalling and diving, the only other possibility is that you threw the model upwards, or excessively hard (and caused a stall) or not hard enough (and it fell to the ground). An unwanted turning tendency can be traced back to a non-central rudder (cripes, how did that slip through your pre-flight checks?), out-of-balance wings, or a warp. If you cannot see the problem, or if the unwanted turn is only slight, move the rudder trim lever to the opposite side and test-glide again until the model glides straight.

If your test-glide goes well, don't keep repeating it. A well-trimmed model is safest when it is high in the sky; all models are in danger when close to the ground, as any piloting error leads instantly to a crash. If, on the other hand, you find a serious problem, it makes sense to keep test-gliding until the model behaves itself. Better to crash from shoulder height than from hundreds of feet up.

SURVIVING THE FIRST FEW MINUTES

The plan for a successful first flight will sound boring, but you can bet that it will be the most nerve-racking few minutes of your life. Your task is to fly straight into wind, then drift alternately to left and right until one of several things happens: your nerve may give out and you have to attempt a landing; the model may drift too far in either direction and land; it may drift downwind and land; it might even drift to the bottom of the hill. Far more likely than any of these, however, is that you will push the stick the wrong way, and the model will crash. That's what happened on my first flight, and on the first attempt of almost every beginner I have watched. Don't be concerned – we all do it. Here's the procedure in detail:

With the model safely launched (I'm coming back to that, remember), allow the model to fly straight into the wind. You can leave it on this flight path for quite a long time – a minute or two at least – so there's no need to worry about changing course. Initially your task is to observe the model closely and check any unwanted tendency. The basic trim should be reasonably good after your test-glides, so any problems should be minor. The elevator trim is the first thing to get right. Ideally the model will fly with its nose inclined slightly downward, and will penetrate forward into the wind at a slow, steady rate. If the slope is producing lift, the model will rise. If the model picks up speed and heads for the valley floor, move the elevator trim lever back towards you slightly. If the model rises rather steeply, loses speed and starts flying backwards, move the elevator trim lever forwards (away from you). With the elevator trimmed correctly, you are half-way to success.

It is very likely that the model will veer slightly to one side. Even a perfectly trimmed model will start turning if you launch slightly to one side of the wind. Gently move the rudder stick to the other side to counteract the turn, and hold it there until the model responds. If it keeps on turning, push the stick further, but never in a jerky manner. Once the model's nose has started swinging back into the wind, ease off the rudder movement. From now on move the rudder stick in tiny increments to keep the model pointing straight away from you. If you find that it constantly tries to turn in one direction, that is the time to move the rudder trim lever until the tendency is cancelled out. If you keep your nerve, you may well end up with a model which flies in a perfectly straight line. Congratulations! It took me about ten flights and three repair sessions to get my first model to this stage.

If you allow your model to continue on this course, it will disappear over the horizon, so we now have to concentrate on keeping it in view. To do this, apply a little left rudder and hold it on until the model's nose swings slightly to one side. "Slightly" is the important word here. If you kick the model into a steep bank, not only will it turn sharply, but its nose will drop, speed will build up, and you will be in deep trouble. Once the model's nose is about 10° to the left of "straight ahead", return the rudder stick to centre and watch what happens. The model should drift across the wind to the left, its nose still pointing almost into wind. If it straightens up again, apply more left rudder. If the nose swings right round, so that the model starts flying downwind toward you, apply right rudder swiftly to straighten up into wind, then try again. The ideal state is for the model to "crab" sideways across the wind, without running downwind at all. When the model has travelled a hundred yards or so, you have to turn right to bring the model back into wind. Straighten up by holding on slight right rudder, and concentrate on holding the model's nose straight into wind again. When you are happy, apply more right rudder to swing the model's nose slightly to the right, then centre the stick. The model should now "crab" sideways again back to centre, then off to the right. When you reach a reasonable distance on the right, turn into wind again, and repeat the whole performance. Essentially the model is flying a zig-zag course into wind, but because the wind is blowing it back towards the hill all the time, the model's path relative to the ground is more or less a straight line to left, followed by a straight line to right.

One important point here: all the turns I am describing are very gentle, the angle of bank being no more than about 5°. Each turn will take many long seconds to complete, but there is no hurry. When you are more proficient, you will be able to make these turns as sharp as you like, but by that time you will have learned that a sharp turn requires the use of the elevator as well as the rudder, and in the very first stages the last thing we need is a complication. If you keep all turns as gentle as possible at first, you do not need to consider the elevator stick at all. A colleague of mine made all his first attempts at slope flying on his ownsome, and each flight came to an early end with a high-speed landing half-way down the slope. He was practising the standard to-and-fro flying pattern in the correct way, but found that his model lost so much height at each turn that he quickly ran out of height. Later on, when he had developed into a pilot of considerable ability, he complained that the magazines and books did not mention that up-elevator was needed for all turns. Well, there's some truth in this, but I believe that where the specialist literature failed him was in not stressing how wide the turns had to be kept. Virtually any rudder/elevator model will turn sweetly without the need for elevator, provided that the turns are kept really wide and sweeping.

If you can keep this course going for a few minutes, you are a natural flyer and can immediately move on to the section on landing. If not (my first attempt lasted about twenty seconds), one of the following disasters will occur:

You never get the model on a straight course into wind in the first place. Never mind; recover the model, carry out any repairs required, and have another go. Holding a straight path sounds so easy, but it is anything but; many models appear at first to be totally incapable of holding a straight line into wind. Never fear: they all learn the knack in the end.

You manage the first bit, and the model climbs straight into the distance, your first turn to the left takes a while to happen, so you bang the rudder over too far, and the model banks into a circle. From that moment your fate is sealed: you desperately waggle the rudder stick about (and probably the elevator too) in a desperate attempt to resume straight flight. The resultant crash may or may not be serious, but it is inevitable. Moral: be gentle with the rudder stick, and counteract any turn immediately it starts.

You get the model drifting to one side, but the attempt to turn it back towards you results in uncontrolled circling or a spiral dive, a drift downwind, and a crash. Other possibilities are: too little lift, and your model drifts to the bottom. Too much wind, and your model disappears back over your head.

Please don't think I'm being negative – we all started like this, and did more repairing than flying in the early stages. You'll soon get the hang of it. Unless you are very unlucky, you will be able to fly from left to right and back again several times without crashing after only a few attempts at maintaining this basic pattern.

LANDING

There are many variations on the theme of landing, but most sites will allow you to land in the following way:

Assuming that you have the standard hill soaring pattern under control, and your model is happily drifting from one side to the other in front of you, all you need to do is allow the model to drift slightly downwind on each cross-wind leg until the ground drifts up to meet it (Fig. 9.8). Unfortunately, if you continue standing at the lip of the slope, this gradual downwind course brings the model back towards you and eventually behind you. Don't let this happen! Instead, walk downwind before starting the landing approach, until you reach a point where you can still see the model and the slope clearly, but have as much flat landing area in front of you as possible. Now you can allow the model to slip downwind gently until it is flying the same to-and-fro course over the landing area. The lift will be much weaker here, and the model will glide lower and lower until it lands. Unless you are very fortunate with your site, and have a vast amount of flat, unobstructed terrain to land on, you will almost certainly have to be careful with your approach in general, and the final turns in particular, as the very wide turns which I have been advocating take up a lot of room – room which might accommodate trees, fences and the like. Just do your best, don't worry too much about damaging the model, and get it down somehow.

Fig. 9.8

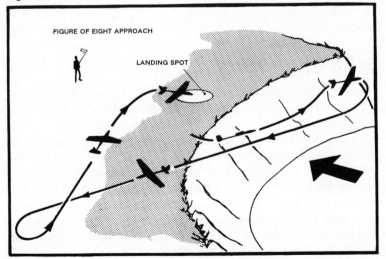

FIGURE OF EIGHT APPROACH

LANDING SPOT

LAUNCHING

Ha – you thought I'd forgotten!

Read all that Dave Jones has to say on launching thermal soarers, then sit back and feel smug. No bungees, no tow-lines, no winches, no gigantic car batteries for us – just a set of arm muscles.

For the vast majority of slope gliders the launching procedure is simplicity itself: switch the radio on, carry out a final check of all functions, then pick up the model in the right hand (left if you're cack-handed). Grasp it firmly at the Centre of Gravity, and check by half-tossing it into the air that the position of your grip will not make the model stall or dive when you hurl it. Hold the transmitter in your other hand, walk to the very edge of the slope face, and hold the model just above your head. Check that the wings are level, and the nose is pointing slightly down from horizontal. Push the model forward once or twice, and feel the wind trying to take the model up and away from you. When you feel confident, simply push the model forward (not up), return your launching hand to the transmitter without delay, and step back a couple of paces, to ensure that the rapidly rising model is in full view.

If the model rises so quickly that it is almost immediately over your head, push the elevator stick forward (away from you) slightly until the plane penetrates into the wind, and is making headway out over the slope. It is very dangerous to try and control your glider when it is low down and directly above you. If the glider won't move, you move instead. It is only when you have an oblique view of the model that you can see what it is doing.

BASIC FLYING SKILLS

Sooner or later you will surprise yourself. You will find that you have returned home from a day's flying with the model un-damaged, and five or six more or less successful flights under your belt. Okay, so a couple of the landings were crashes in all but consequence, and once or twice you panicked and only survived by much luck (call it skill). Nevertheless, you have scrambled your way onto rung number one on the ladder, and will never fall back again. It's just unfortunate that that first rung is so high off the ground!

From now on you will find that progress is rapid – provided that you follow one or two simple guidelines:

- Fly as often as you can (every weekend and preferably weekdays as well), and for as long as you can on each occasion. Fly, fly and fly again until your transmitter needle tells you it's time to quit. In my experience this is the only way to ensure that you reach a basic level of competence and confidence in a short time. Later on you may have less time for flying, and disuse will blunt the edge of your skills, but once you have this basic knowledge and experience it will always be there when you find the time to take up flying again.

- Stick to one model, or perhaps two (to allow you to cope with a wider variety of conditions) for as long as you can. All models are different, and each new model takes quite a long time to get to know. I'm sure you will find that your second model will rob you of much of your confidence, as its control responses will be different from those of your initial trainer. You will then have to work hard to regain your former level of skill. If you constantly switch models, you will spend all your time becoming acquainted with them, and will find it hard to improve your flying.

ASPECTS OF FLYING

Practise flying safely. One bad accident caused by you and your model will destroy your enjoyment of the hobby for ever. The over-riding necessity here is to avoid anything which could endanger another person. Flying straight towards spectators or fellow modellers, showing off with low passes along a crowded slope, flying low over cars, horses, all these must be avoided at all cost. Similarly important is the condition of your aircraft and radio. Never fly if you are unsure of the state of charge of your batteries, for example. If there is a crease in your foam wing after a wingtip-first landing, take it home and investigate; a folded wing in the air turns a docile glider into a high-speed missile.

Practise the correct use of the elevator. We deliberately left this control alone at first, but you cannot avoid it for long. It is a glider's speed control. To go faster, push the stick forward. To slow down, pull the stick back. At the same time, the model will dive and climb respectively, but it is dangerous to think of the elevator as an up/down control. Basically, we all depend on the lift generated by the hill to take the model up. If you want to come down to land, all you do is fly out of the lift area, and the model will come down safely. If you wish to land by holding in down-elevator – well, try it once when I'm not around if you wish, but I don't think you will try it a second time. For most turns, apart from the very gentlest variety, you need to apply slight up-elevator to prevent the nose dropping; left alone, the turn will develop into a spiral dive. Gliders vary greatly in this respect; some circle smoothly and sweetly with little or no elevator control, others need constant elevator adjustment to hold a level turn. Investigate your model's turning behaviour until you have mastered it under all conditions: how much rudder to apply to initiate the turn, how much (if any) to hold on in order to keep the model turning, and how much elevator to use. To become thoroughly familiar with any model takes hours rather than minutes – don't rush it.

Practise flying smoothly. With the possible exception of its final one, you never see a full-size aircraft carry out a jerky manoeuvre. There are times when you will need to whack the stick firmly into one corner to avoid a mid-air collision or rampant oak, but generally speaking all stick movements should be unhurried and gradual, including the movements back to centre. A sure sign of a nervous beginner is the continual click heard when the stick is released and snaps back to neutral. Moving the sticks smoothly requires a definite effort of will; many pilots watch their model intently, without having any real idea what their fingers are doing. Every now and again try watching your stick fingers doing their thing, and ask yourself if their movements could be smoothed out somewhat. Watch other pilots, and compare the hurried, nervous fidgeting of the "unlucky" flier with the confident progressive handling of the expert.

Practise flying accurately. Most of the time you can get away with flying in roughly the right direction, but if you have the ability to position your model exactly where you wish, you will have a definite advantage in a tight spot. One day you will be flying happily on your own when a group of visitors or clubmates turns up, and suddenly the sky is full of aeroplanes. This can be hard on the nerves for the relative beginner, but if you can position your model confidently to one side of the others, or above them, or land quickly and accurately, you will find your own way out of the trouble. There are many ways of practising accuracy. When the slope is all yours, try flying your model cross-wind exactly one wingspan away from your transmitter aerial's tip, or exactly at ridge-height, or precisely along the horizon, or along an imagined line from your nose to a tree down in the valley, or between two bushes on a (high) landing approach. The skills you acquire in this way will never be wasted. Another good ploy is to practise loops from a given position, diving into wind and starting the loop exactly when the model crosses a house on the horizon. Complete the loop and recover at the same point. Endlessly difficult, but loads of fun, and time well spent.

Practise flying purposefully. It is all too easy to get into a rut with any form of model flying, particularly if you are stuck with one model, but you can avoid it easily by setting yourself tasks. Go out one day and concentrate single-mindedly on flying concentric circles in front of you. You'll find it darned hard at first, but eventually it will become second nature. If you find a particular flight manoeuvre difficult or impossible, work out a plan for mastering it, and plug away until the skill is mastered. Inverted flight, for example, can be started with a simple roll; then a roll with a hesitation at the inverted stage; then the inverted stage held a little longer with a touch of down-elevator; eventually a protracted period of straight inverted flight; finally perhaps an inverted circle. A whole session devoted to tight pylon turns, concentrating on minimum speed loss, will do you the world of good. In fact, every flight manoeuvre will benefit from concentrated practice, and you will feel distinctly virtuous once you have made progress in this way.

Practise flying in a wide variety of conditions. Your first model will probably only be happy in a narrow band of wind speeds, but once it has served its purpose (taught you to fly), why not try it in a slightly harder blow just to see what it can handle? There is nothing to lose and everything to gain by flying in what appears to be too little breeze. Your endless hours of practice at smooth, accurate flying will enable you to fly your model "on tip-toe", skating round the very rim of the bowl, lapping up every scrap of lift that's on offer. To my mind this is one of the most enjoyable and challenging forms of slope flying of all, and is available to everyone. For any model there is a minimum quantity of lift below which it will fall out of the sky, but only 1 mph above that figure, and you can enjoy the fun. Ballasting gliders is not really a subject for a beginner's guide, but suffice it to say that adding a lot of ballast (don't mess about with 2 ozs. – make it half a pound at a time) can markedly increase a model's ability to penetrate a blustery wind. Personally I'd sooner shift the radio into a more suitable model than add ballast, but then perhaps I've seen too many crashes which were directly related to excess ballast.

Practise simple aerobatics. Loops are the simplest man-oeuvre of all, and make the beginner feel great. Fly the model into wind at a good height, and well out in front of you. Push the elevator stick forward gradually until the model has picked up lots of speed, and is diving at about 30°. Now gradually return the elevator stick to centre, and pull it back towards you. The model should climb, turn on its back, then dive down and resume level flight. At this point, return the stick to neutral. If the model doesn't complete the loop, you didn't dive long or steeply enough. If it turns a peculiar somersault at the top, you applied too much up-elevator. If it does two loops in a row, you forgot to neutralize the stick. A spiral dive is fun to watch, especially once you know that the model will pull out of it. With the model at great height, gradually apply full right rudder and up-elevator, and see what happens. When you feel your heart climbing up your throat, centralize both controls and the model will pull out by itself. Don't yank in up-elevator in panic, otherwise the wings will probably part company from the fuselage. A stall turn is difficult to accomplish with most rudder/elevator gliders, but is worth trying nonetheless. Flying crosswind from left to right, apply down elevator until the model is diving steadily. Slowly pull in up-elevator until the model is climbing vertically. Release the elevator stick, and bang the rudder stick over to the left just before the model loses all its forward (upward) speed. The model might, just might, yaw round through 180° and end up falling vertically again. Pull in up-elevator smoothly to recover. You are bound to get the timing wrong at first, with the result a gigantic, highly entertaining stall, but never mind. The manoeuvre is actually much easier to do in a recognizable fashion with a three-axis (aileron, rudder, elevator) model.

LANDINGS AGAIN

Above all, practise landings. So many modellers relegate them- selves to being forever second-rate pilots because they are scared stiff of landings. Don't let it happen! Landings can be just as much fun as any other flight manoeuvre, provided that you don't avoid them. In theory the slope flier can have a day's flying with only a single landing, but I would strongly advocate going to the other extreme: once you have a model which you can consider expendable – when you have built and successfully trimmed your second model, for example – you might like to go out with your old banger and fly nothing but landings. Launch, climb to fifty feet, then circle and land. Launch, circle, and land. Keep going until your batteries are flat. If you are really well

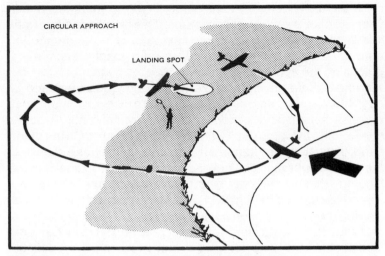

Fig. 9.9

acquainted with your model – and you should be by now – you will soon find that you can land safely every time. After that, you must be sure to take a hat with you when you go flying. Place the hat on a suitable divot, and aim to land on that each time. Keep a count of your success rate, and turn the activity into a game. I've spent countless enjoyable hours trying to land on a particular daisy; it really is immense fun, and the skill is so very useful.

I have mentioned the simplest form of landing already, in which the model is kept in front of you all the while. Once you have mastered that, the next step is to practise flying towards you, as all other landing approaches involve bringing the model round behind you and flying forward, into wind, towards your feet. It will be apparent that applying right rudder to a model which is heading for you will cause it to turn to its right, but to your left. This is exceptionally confusing at first, and is yet another manoeuvre which you must practise again and again. Some people advocate looking over your shoulder at the model, others swear by "pushing the stick under the low wing"; try it by all means, but in my opinion there is no artificial answer. The only way to be in control at all times is to keep on doing the manoeuvre, preferably in a safe position initially, until you never need to think which way to push the sticks.

The most common landing approach with slope gliders is simply to fly a broad circle starting from an upwind position, bringing the model round behind you and into wind again (Fig. 9.9). The model is then guided in a straight line onto the ground. This is perfectly satisfactory in most situations, and if your site allows such an approach, practise again and again until it is second nature. Many pilots have a built-in preference for right-hand circles over left-hand, and I strongly recommend practising the approach from both sides, as sooner or later you will find yourself flying at a site where only one approach is possible. A square approach (Fig. 9.10) is a nice refinement, and is widely held to be good form, but it is only mandatory in scale competition flying, so don't worry too much about it. On the other hand, if you have any ideas of flying scale sailplanes, practising the square approach can never start too early. The final form of landing which is the only type possible at many sites, is the slope-side approach (Fig. 9.11). Some slope sites have no real landing area at all to offer, but are just so good in other respects that you have to make do. The answer here is to land on the slope face itself. This is quite daunting at first, as it can be very difficult to reduce the aircraft's speed enough to give a reasonable chance of a soft touch-down. The trick is to deliberately lose height by continuously looping, rolling, spinning or spiralling until the model is quite a long way below the rim of the hill. It can then be gently flown up again, flying in the standard figure-of-eight pattern along the face of the hill, until it is nearly at the top. With the model very close to the ground, and the wings banked parallel to the incline, turn gently up the slope, easing in up-elevator as you do so, so that the model slides sideways up the

Fig. 9.10

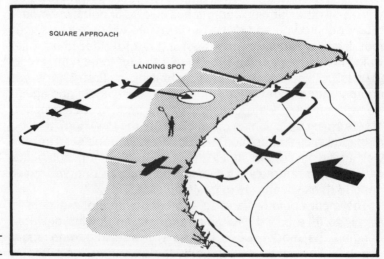

SQUARE APPROACH

LANDING SPOT

hill and rests its nose on your boot. If this sounds difficult, well, it is difficult, but then you can spend all day practising it, can't you?

HIGHER THINGS

When you have worn out model number 2, it is time to consider a model with ailerons. A small aerobatic model (often known by the appalling generic name of "kipper") (Fig. 9.3) may suit you if you have learned to fly rapidly, and have a lot of confidence in your ability. If not, perhaps a larger, semi-scale model with all three principal controls would be a better bet (Fig. 9.4). In either case, be prepared for a set-back. You have spent many hours learning how to turn a model with the rudder, and you are about to discover that ailerons respond entirely differently. On a glider with plenty of dihedral, the rudder only turns the model in-directly, in conjunction with the geometry of the wings. The rudder does not bank the model directly. Generally speaking, to come out of a rudder-induced turn, all you need to do is neutralize the rudder stick. In contrast, the ailerons directly effect a change in bank, but do not necessarily initiate a turn on their own. Push the rudder stick to one side when flying a rudder/elevator model, and after some delay the model banks over and starts turning. If you change to an aileron model and push the aileron stick to one side, the model banks immediately,

Fig. 9.11

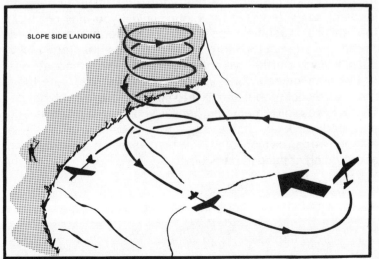
SLOPE SIDE LANDING

but does not turn, instead flying in a straight line for a few moments before scything downward in the direct of the low wing. You need to apply the ailerons smoothly to set the angle of bank, then neutralize them, and apply slight up-elevator to sweep the model round in the turn. You will often need to apply opposite aileron to get the model to recover from the turn. Once you have mastered the art of turning with ailerons, the whole repertoire of aerobatic flying is open to you. You may well find that loops are more difficult than with your trainer, but in compensation stall turns are much easier, and all the rolling manoeuvres become possible. Once you have reached this stage, you are a beginner no longer.

From this point on you have little to fear. Every time you go out and fly you can try something new, and you will find the old manoeuvres get easier and easier. Rolls soon become so easy that you will flick through them without a thought. The initial panic when the model is inverted soon fades away, and you will find yourself flying circuits and even overshoots with the fin pointing down. The only danger now is that your imagination will fail you, and you will get bored with the same old rolls and loops. If this should happen, it's a good time to send away for details of the BARCS slope award scheme.

This scheme is intended to raise the overall level of competence of slope fliers, and does just that in a most enjoyable way. Work steadily through the set programme of manoeuvres and achievements and your flying skills are bound to improve. At the same time you earn some pretty stickers for your models and loads of self-esteem. At the higher levels you may not have thought of trying some of the aerobatic manoeuvres prescribed. At the first attempt the more complex figures are pretty tricky, but plug away and they will come right in the end. If there are one or two manoeuvres which you just can't manage, write them down on a sticky label and fix it to your transmitter sub-shade. That way you will remember what it is you have to practise. You may find, as I do, that a day spent polishing up a single manoeuvre can be more enjoyable than a session spent messing about doing nothing in particular.

10   FOUL WEATHER FLYING

Britain could never be said to have a monotonous climate: wind rain, hail, snow, Arctic freezes and Saharan blasts, you name it and we get it. This is, of course, not good news for an outdoor pursuit like soaring. Not all that many years ago we would have all packed up and gone home on half the days that we now fly on. In competitions the need to extend the performance of models to withstand and perform well in poor conditions has led to dramatic increases in their abilities to withstand foul weather. All this has led to the use of new flying techniques and procedures.

WINDY WEATHER

The first thing to do is define the term "windy". For most of our purposes let's assume that for thermal soaring that means over 15 mph and up to about 30. Anything over 30 mph may be possible for the model but it is tiresome for the pilot and not a lot of fun anyway. There is a complication here, as the wind speeds given by the weather forecast are generally accurate and are taken at a height of thirty feet in a flat unobstructed area. The readings that we get from a wind meter (anemometer) held at some 7 ft. above ground level are affected by the surrounding countryside to a large degree. Whatever the wind meter reads add 50% for the windspeed at height, thus 16 mph at 7 ft. is more likely to be some 24 mph at anything over 30 ft. So consider it windy once the wind meter starts to read 10 to 12 mph.

The main consideration is can the model take it? If you have any doubts pack up and go home. The best way to find a model's limits is to fly it in gradually increased windspeeds, not all of a sudden in a stiff breeze. The two major worries are will it be capable of standing the launch stresses and will it be able to penetrate forward in the wind? The first is a structural problem, the second involves aerodynamics and design.

Generally speaking if the wings (not the joiners) bend a lot under tow in normal conditions then they are unlikely to withstand high wind towing, which can be seen on the more lightly built models commonly referred to as "floaters". If the wings are stiff under tow then they should be satisfactory up to about 25 mph (15–17 at ground level). Beyond this sort of wind speed they are likely to need careful launching, and this is the sort of maximum windspeed that a current everyday Open Class model can be flown in. Over 25 mph you need to go to a much tougher model altogether, something like a competition 2 Metre or F3B machine. Small models are generally tougher than large ones and more capable of standing the extra stresses, and in addition their lower overall drag makes them more suited to high wind flying. A strong 2 Metre competition model will fly in almost any conditions if ballasted properly.

The wings for high winds need to be torsionally stiff as well as strong in bending. Torsional stiffness (resistance to twisting) will prevent the flutter and "tuck-under" problems that normally kill off a lightweight soarer that is flown above its natural speed range. Try twisting the wing gently at its wing tip, if it feels stiff and springy, somewhat like a good foam wing, then it should be satisfactory, but if it takes little force to twist the wing then it is likely to flutter. If you find that most of the twist happens at the root end of the panel then tuck-under or flutter are likely to occur at quite low speeds. If the model is known to be capable of high speeds in the dive then it will probably be stiff enough. The covering material makes a big difference. Tissue adds torsional stiffness, as does glass cloth.

Whether a model will penetrate in a wind depends upon the

An example of a lead-filled ballast tube for wing mounting. Needless to say, ballast should not significantly affect balance.

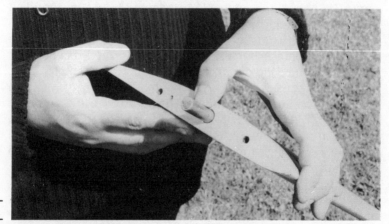

wing section used, the total drag of the model and its wing loading. The sections used on most of the newer designs are generally faster than the ones in common use in the late 70s. They were frequently proved in F3B, where a wide speed range is an essential quality, common examples being Eppler 193, Eppler 205, the Quabeck series, and the fast Selig sections such as S.3021. These are low camber medium lift sections and they can be recognised by their ability to keep accelerating as more and more down trim is fed in. The older high lift sections are superior on a calm day and in light lift, but soon run out of speed range as the wind rises. To produce high lift they use more camber, which produces more drag at speed. The effect of this can be seen if down trim is fed in, when the model will accelerate to a certain velocity and will only go faster with a lot of down applied, and a lot of height burnt off.

Low total drag of the model will allow it to make easier headway. The wing accounts for some 90% of the drag, but a smooth streamlined overall shape will help to keep the drag to a minimum. A boxy model with banded on wings is less likely to be able to penetrate.

The total drag of the model, if low will allow it to make easier headway. The wing accounts for some 90% of the drag, but a happier in the wind. The determining factor is the wing loading, which should be raised as the wind rises. The process is of course ballasting, and each model will have its own ballast requirements. Generally speaking there is little point in ballasting a model with high lift section, as once it has hit its high drag speed it will just come down quickly.

Ballasting is an art in itself and it will take time and experience before the best ballast for the prevailing conditions can be found. It is no good increasing the wing loading by much less than 2 ozs/sq. ft. at a time and thus a 100s model will need some 11 ozs. of lead for each increment of ballast. There is a practice of putting some extra lead in the nose on windy days, which is only of assistance if you fly with a rearward C.G. position. The effect is akin to putting extra down trim in the aircraft and it does not make it any more capable of speed. Having gone to all the trouble of trimming the aircraft it is pointless altering its known handling characteristics just because the wind is blowing.

A reasonable estimate of the amount of ballast required can be gauged by holding the model into wind and gently loosening your grip. If it lifts and tries to pull up and out of your hand then it is probably too light. If it lies there heavy and docile and does not want to lift then the ballast is probably too much. If the model lies docile during the lulls in the wind and starts to go light in your hand during the gusts then it should have enough balllast for the conditions. This method should give you a model that will stay stationary during the gusts and penetrate gently the rest of the time. It is also better than launching without some idea of the results. The Beaufort Wind Scale and recommendations in Appendix 1 may be useful to you.

Ballast will make the model handle differently. The airspeed and the stalling speed will be greater. As a result the turns will have to be flown faster or on a greater radius to avoid stalling out. The extra inertia of the ballast will make the model feel ponderous, yet the controls will be responsive if the airspeed is kept up. While the model will respond to the controls, the recovery from violent manoeuvres may be slow due to the inertia effects; the best policy is to fly fast and smooth and avoid any steeply banked or energetic manoeuvres.

Flying in a high wind poses several specific problems: thermalling, staying in the field, landing approaches and landing. Thermalling in terms of circling in the lift is a problem. The trouble is that you will be very rapidly carried down wind, which is acceptable if the model can penetrate back upwind easily but not a good idea if it leads to an out-landing. The only other course open to you is to "pole bend" as mentioned in Chapter 8.

Staying in the field can be awkward if the ballast is inadequate, since the model will be slowly but inexorably blown back. The best bet is to land as quickly as possible – put the nose down and force the model forward by converting height rapidly into forward speed. If airbrakes are fitted and they are not too large try deploying them as well, which will increase the effective wing loading of the model and may well allow it to come forward, albeit with a great loss of height. If the model does get blown away try to put it down where you can see it, for if it goes out of sight the results could be nasty to the model or someone else.

If the ballast is correct then the model should be making headway upwind and the problem here is turning back down wind for landing. The groundspeed (the speed that the model is travelling in relation to the ground, and you) of the model will be dramatically different into and with the wind. Look at it this way; windspeed 20 mph, model's airspeed 25 mph, and if the model is flying into wind the groundspeed will be the airspeed minus the windspeed, i.e. 25–20 or 5 mph. If the model is flying down wind the ground speed will be the sum of the airspeed and the

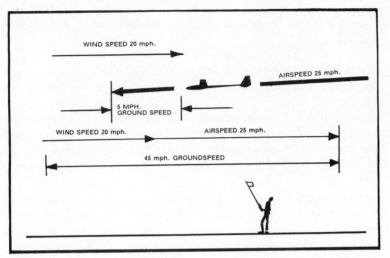

Fig. 10.1

windspeed i.e. 25+20 or 45 mph, as in Fig. 10.1. Two problems occur, both pilot-induced. Firstly when turning downwind the pilot will tend to underestimate how far the model will travel on the downwind leg and end up a long way back before the model comes back into wind, possibly out of the field. If you are going to turn and perform a landing approach in the standard manner do so from well up wind and turn in crosswind when the model is at 90° to the wind, directly to one side of the flight line as in

Fig. 10.2

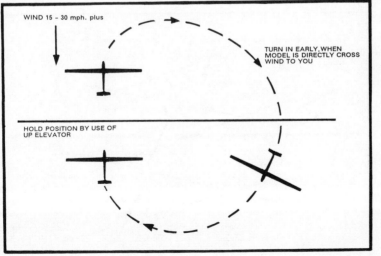

Fig. 10.2. The chances are that it will then end up where you want it to be.

The other pilot induced problem with flying downwind is that the natural groundspeed and sink rate of the model will look excessive and the natural response will be to hold some up elevator in to slow it down. A bad mistake this, the results of which are only too predictable. The next thing that the flier tries to do is turn the model into wind, but the up elevator has robbed the model of airspeed and the rudder has lost steerage power, result, very sluggish response or none at all. Probable outcomes either an out-landing or a downwind landing at a ridiculous groundspeed. Likely result damage to the model and anything else in the way. This is a classic pilot error on the slope and has led to the "always turn into wind" dictum. It is far better to feed a little extra down in on the downwind leg in order to maintain control response, and ignore the height loss.

The obvious question is what do you do if the model is not far enough forward or high enough to turn back normally? The answer is to fly as if you were on the slope. Fly across in front of yourself and then back again the other way ("essing"). As you do this start to fly the model at 45 to 60 ° to the wind, as in Fig. 10.3, which will allow the model to drift back downwind without ever actually turning downwind. Once the model is in a favourable position point it back into wind and carry on. Flying in a wind is always tricky so practise these manoeuvres at height to get used to them.

Fig. 10.3

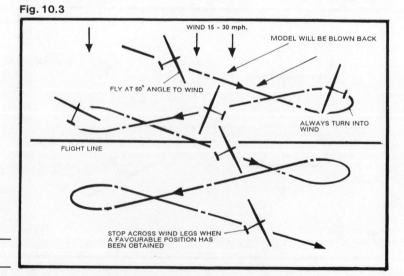

Another solution, albeit one that requires confidence in and knowledge of the model, is to ease back on the elevator and lift the nose of the model. This will do two things: since you will be flying with a fair amount of down trim in you will be flying well away from the stall, the up elevator will bring it close to the stall and increase the lift. The model will rear up, rise and go backwards. As the angle of attack rises so does the drag and this will help it to be blown back. The danger of course is getting too close to the stall, particularly if this is done close to the ground. Try this out at height and practise the recovery. Fig. 10.4 shows the idea.

A probable natural problem will be turbulence, but you can always tell if it is likely by looking at the edge of the field. If it is surrounded by houses or trees beware. If it is a low hedge or fence it will still be present but not so marked. If you hit turbulence at height you have a chance of recovering control, but if it hits you at low altitude you may well lose it. The answer to this one is airspeed, plenty of it. Airspeed equals control and given control you have a good chance of fighting your way out of trouble. So when you are landing do so with plenty of airspeed – the ground speed will probably be low so the potential for damage due to landing speed is low.

A good way of dealing with turbulent air is to drop through it quickly using the airbrakes. The procedure is to do your approach but come in a bit high. Steady the model, deploy the brakes and do a shallow dive at the same time. Level out and

Fig. 10.4

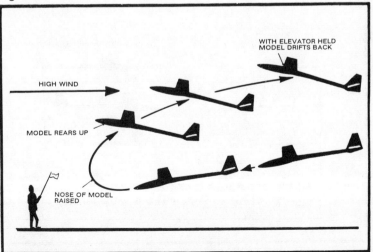

WITH ELEVATOR HELD
MODEL DRIFTS BACK

HIGH WIND

MODEL REARS UP

NOSE OF MODEL
RAISED

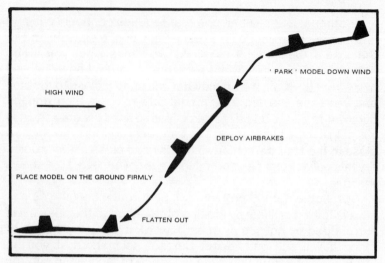

Fig. 10.5

place the model on the ground as before. This requires practice but is a safe and fairly precise method. If you try to flare out and land in the normal manner you are very likely to get hit by turbulence when you are at low speed and unable to respond quickly. See Fig. 10.5.

To further complicate matters there is wind shear to worry about. The false windmeter reading problem is caused by it. Imagine a stream of steadily flowing water and stir up the mud.

Fig. 10.6

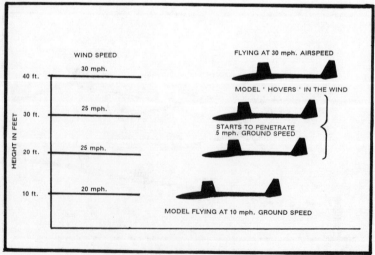

The mud at the top will move fast, the lower down slower and that close to the bottom will move slower still. It's boundary layer effect, arising from the air close to the ground having to flow over the ground's surface and in so doing being slowed by it. The air above this layer slides over the slowed lower air more easily and can travel faster. The air above this layer can go faster still and so forth until the clear air windspeed is reached. Try lying down on a windy day to see the power of the effect.

Wind shear will affect the model as it lands. It may hover at about 30ft. but as it drops it will accelerate forward as it moves through the slower moving layers. The problem that this poses is that it is difficult to gauge where the model will land. You may well think that it is too far downwind only to find that it makes you jump out of the way as it flies upwind to land in front of you. This is not too problematic unless you are trying to get landing points in a competition, and the only answer to this one is practice and knowing your model. Fig. 10.6 may convey the basic concept better than words.

RAIN

Sooner or later you get caught in it, usually at a competition, often in the middle of a sport flight. It does not pose insoluble problems and is a minor annoyance most of the time. The model seems little affected by rain and indeed I have encountered good lift and seen many competition flights go full length despite the downpour. The problems are for the flyer. Discomfort is the worst part, since you know that you are stuck out in the weather till you land and this is when you appreciate good flying clothes. It is difficult to fly facing into the wind, particularly if you wear glasses. The answer is to fly to either side of the flightline or better still downwind so that you can turn away from the wind and rain.

Check that the wet is not getting into your model's radio gear. Short (up to 2hr) exposures seem all right but continual exposure to damp atmosphere can give glitches and radio failures. Put the fuselage in the car and dry it out if in doubt. Tissue covering suffers badly in these sort of conditions and plastic films and glass cloth are much better. The transmitter should not be exposed to the rain if at all possible, as it gets through the slots for the trims and the stick gimbals and causes shorts in the trim pots and other items in the vicinity. The usual symptom is for the model to suddenly set its neutrals to an extreme position, three quarters of rudder or elevator. Not a lot of fun if it happens in flight! Leaving the transmitter in a warm room to dry naturally seems the best bet when you get home.

At the field there are a few items of good practice to note. If rain is likely leave the transmitter lying face down, as the back of the transmitter is less vulnerable. A perspex rainshield provides good protection, as does a carrying box. An old dodge is to poke a hole in a polythene bag and slip the bag over the aerial and transmitter. Your hands are then put in the bag and the model flown in the normal manner. It can make a good windshield to keep your hands warm on the slope, too!

Often the model is covered with plastic sheet while it is raining, but beware the weight of water that puddles on the sheet. The tailplane can get bent down and damaged if you do not either remove it or use some wire hoops to keep the sheet from making contact. Wings are usually strong enough. When you get home let a soggy model dry slowly, dry the outside and leave the parts standing with no pressure on them for a day or more.

Keith Thomas informs me that a popular alternative is to leave the rigged model under the car, though this is not so feasible with a big thermal soarer. However, there can be a problem, "Two hours later you give up and drive off . . . crunch!"

LOW CLOUD

This can be encountered on the slope or occasionally on launch from the flat and is a danger when thermalling. As soon as you enter low cloud apply full up elevator and full rudder, which puts the model into a spiral dive. Now look behind the spot where the model disappeared and to the side that you turned towards, i.e. if you turned to the right look some thirty yards to the right and some fifty yards behind the last known location. Don't panic – wait for the model to reappear and recover from the spiral dive, using the airbrakes if you have a lightweight wing to worry about. It is very useful to have someone else looking as well since they will probably be less nervous and more analytical about the situation.

POOR VISIBILITY

Heavy haze or mist are the usual culprits and it is very easy to lose sight of the model under these conditions. I well remember flying the wrong model for some time in one competition, and it took a long time to repair the damage. You usually lose contact when the model is flying directly towards or away from you and its frontal or rear area are all you have to see. The best procedure that I have found is to not let the model fly in a straight line under these circumstances. Fly gently from one bank to the other so as to expose some side area, or approach the desired position from an indirect and more visible course. If at all possible do not fly the model too far away from yourself. A dark colour scheme makes life a lot easier.

COLD WEATHER

This poses a few problems. The plastic components in the model may well become less compliant or even brittle, so that hard landings are to be avoided as the model may shatter something. Tape goes hard and non-sticky unless you keep it in your pocket till it's needed and apply it straight off the roll. The flying clothes will really be tested, particularly the gloves. The battery charge may be affected, so play a little safer on flying time. Snow is not a problem at the field but make sure that it is cleared out of the equipment bay before the model is brought into the house.

You may think that you will never fly in such conditions, but if you have been cooped up for a fortnight waiting for a good day you are more than likely to risk less than ideal weather. If you fly competitions you will almost certainly fly in all of the above at one time or another. The experience will make you a more capable flyer.

11 THERMAL SOARING COMPETITIONS

WHY BOTHER?

There seem to be three broad types of glider guider, although each type may fall into one category to a greater or lesser degree: the sport flier, the competitive sport flier, and the competitive flier. A newcomer will inevitably start off as a sport flier and how his personal philosophy affects his (or her) attitude to model flying will determine whether he becomes competitive or not. What I am trying to say here is that competition is not for everyone. Personally I envy those happy souls who can appear at the flying field with a bog standard kit model and fly just for the soul-liberating pleasure of seeing their handiwork in the air. Let's face it, we all suffer the pressures of modern life, so why add another one to the list? So those of you who feel that you are inveterate sport fliers read no further.

So why does one become a competitive flier? It seems to me that the sort of person who gets involved in competitions requires new challenges, feels the need to test his abilities against others, does not like his skills to remain undeveloped, enjoys the competition camaraderie, and above all else likes to find and extend his personal limits. Without these or very similar motivations it is pointless becoming involved in competitions. The obvious line of development usually goes like this: Modeller builds first model and, if lucky and well instructed learns how to steer it round the sky. He then becomes proficient enough not to endanger the club's insurance policy and starts to fly solo. Pretty soon he realises the limitations of the training model (if it hasn't already found a permanent resting place) and he then builds "the second model". This is frequently a 100S type glider of far greater performance than the trainer. Once this performance has been mastered and the rudiments of thermalling learnt the developing flier faces a dilemma; the learning curve has flattened out and the challenges are becoming fewer, where does he go

Pete Barnett with his *Little Scratcher*, typifying current 100S size and proportions.

now? There are of course several answers, such as bigger and better models, different types of model, different areas of flying (slope or scale for instance), and lastly the usual answer of contest involvement.

So what does competition do for you? Like any activity the

A 2 Metre model, Chris Tomkins' *Metric 2000*, shows the difference in size from a 100S.

benefits are in direct proportion to the amount of effort and interest that you are willing to invest. The real advantage for those who require it is the constant challenge to one's abilities. The advantage for any flier is that it makes you fly your model at 100% of your ability, whereas sport flying for the experienced pilot may only demand some 50 to 70% of one's attention. The point that needs to be made here is that it is very easy to settle for a level of skill that is adequate, the danger being that one's flying may then quickly become boring, the ultimate result being a loss of interest. To see the benefits in terms of skill all you need to do is watch any group of glider flyers. The ones that seem to be able to find lift in the most marginal conditions and manage respectable flight times will almost without exception have competition experience. One notable local comp. flier is reckoned to have some hidden compartment in his car for the day's quota of lift, which always seems to be around when he's there and seems to leave with him. Thus it can be said with some confidence that nothing else will develop and hone the skills of a newly-competent flier as much as competition involvement.

For those who enjoy designing, competition provides the ideal proving ground for one's creations. It is all very well saying that one's new model flies very well but its merits can only be properly judged by direct flying comparison with other models

You can't get much flatter than an aerodrome for flat field thermal flying. Model rejoices in the name *Viking Bitsa*.

under circumstances where everyone is trying their hardest. In addition the drive for increased performance puts the competition model and modeller in the forefront of the development race. It is difficult to think of any worthwhile or major improvement that has not come about because of competitive pressures – better sections, improved structural knowledge, better aerodynamics, greater capabilities and so forth. There is a negative side to this, however, in that it can easily happen that the technology becomes so expensive and time-consuming as to make participation in a class somewhat forbidding for the beginner. Fortunately we have sufficient number and variety of classes for most tastes and levels of commitment to be catered for.

GETTING STARTED

The best place to start is in club competitions. Stick to them for one season or so in order to build your confidence and abilities. Different clubs will run various types and combinations of events to suit their local geography and membership, so it's not easy to generalise. Since the larger part of the challenge of thermal soaring is improving duration it's not surprising that the majority of events are time challenges of some sort. It is important that you should not feel embarrassed during the period when you are improving your performance, so these low key, fun comps are ideal. Probably the best model to have is a 100S glider, preferably with airbrakes. The 100S class only has one or two national events arranged for it per year but its popularity as a club competition remains strong. Local rules may differ, but the usual way that they are run is to the rules developed by BARCS (the British Association of Radio Controlled Soarers). These require the flier to achieve the longest possible flight time within a period of 8 minutes, this period being known as a Time Slot, or more frequently as just a "Slot". The scoring is usually done by the "Percentage Slot" scoring system. Under this the person with the highest score is awarded 1000 points and all the rest are scored as a percentage of the slot winner's score. This is an important concept to grasp as it is used in other classes as well, the advantage being that British conditions are so changeable that a simple time task may not give a fair result due to one slot being full of lift and the next full of sink. The big advantage of club 100S events for the aspiring competitor is that not only will he be likely to already own a

Mark Sickling's *Pteranosoarer* **is a fast flier. Span is 149in., root chord 13in., flying weight 9lbs.**

suitable model, but the model specification brings all the fliers to some sort of parity.

One thing that is seldom mentioned is that most competition flying requires a greater or lesser degree of teamwork. It is in the club competitions that you will discover who you can (and cannot) work with. This is important because if you are ever going to take part in national events you will need to have a body of people that you can rely on to assist you, and the more such relationships that you can build the better. However, this must be a reciprocal arrangement and you must be as helpful to them as they are to you.

There are four different functions in a team of fliers: the flier, the spotter, the towman and the timer. Each person should be able to perform every duty except perhaps towing, since this requires a moderate level of fitness. The prime function of the team is to reduce the workload of the flier so that he can fly to the best of his ability: anything else is superfluous. Each flier will have different requirements, so the team must adjust to his temperament and needs. Some require the minimum of information and encouragement, others need to be talked through the whole flight. Generally the more experienced flier requires less information.

The Timer
This is the least demanding of tasks. It used to be the responsibility of the organisers to provide timekeepers, but it is now the pilot's responsibility to ensure that his flight is timed, so it becomes a team task. All that is required is that you start the stopwatches (two are required for safety) when the towline drops off the model, and stop them as soon as the model touches the ground. It is to the credit of the vast majority of fliers that this arrangement does not lead to any noticeable cheating. An additional advantage is that it speeds up the running of the competition, since the timer very probably acts as the spotter or can act as an additional spotter, take over the count-down if asked or combine the timing and spotting tasks if the team is shorthanded.

The Towman
One needs to be moderately fit to tow models in average conditions and quite fit to get a model up in calm weather. All the main classes use hand-towing because it ensures a more or less equal height launch for all the competitors. How to tow models is another subject altogether, but suffice it to say that it is not all that difficult and requires finesse as much as brute strength. It is, however, quite a skilled task and one that requires practice to perfect. Ideally a team should have a minimum of two fit towmen. One thing that needs to be clearly understood is the system of signals that will be given by the flier to the towman and by the towman to the spotter. These are essential for successful launching and will differ from group to group, so you will need to learn them as you go along. A good towman is obviously essential for success and technique is only built up by practice, so while everyone else is waiting for a spare bungee get the towline out and get on up there.

The Spotter
The prime function of a spotter is to be the flier's second pair of eyes, the ones at the back of the head that nature never provided us with. It is a task that requires judgement of a high order since he is acting as adviser to the flier, and is thus best left as a job for experienced fliers. During the flight the spotter tells the flier what the rest of the field is doing, whether they are in

sink or lift, how they are spread out and where. From this information he should be able to recommend a course of action to the flier if he asks for it, and should keep the flier updated about the current state of play if it is relevant. A spotter is thus mainly concerned with the rest of the field but he should also be able to advise the flier if he is in sink or lift, or in any danger. The other main function of a spotter is to give the flier a count-down of the remaining time left in the slot. This is vital in good conditions when overflying or landing too early may be crucial to the results. This function may be performed by the timer. No two competition slots are ever the same, so it's not easy to give specific recommendations. Again it is a function that has to be learnt the hard way.

The Flier

The person around which this whole structure is built has to consider the team as well – it does not pay to upset them. You should always remember that you are in charge of the model and in ultimate control of the flight. Advice is there for the taking only if you either agree with it, or trust the spotter's judgement. If anything goes wrong, and it frequently does, it's no good blaming your team, because the responsibility is wholly yours and yours alone. You are the captain and they are the crew. it pays to try to use the same people all the time, in much the same way that a settled team in any sport is more effective than a scratch team. The spotter is the key figure, followed very closely by the towman, with the timer coming a poor third.

Now the above may sound pretty daunting, but it appears far worse on paper than it does in practice. The above-mentioned teamwork and roles come quite naturally and there really is no other way to organise matters. It can be seen then that being a good flier is not enough and that one must develop one's abilities in the other team roles as well. Towing can be learnt on sport sessions or in club comps. Timing is easy enough and, by observing other people spotting can be picked up and practised at club level. It is not necessary to go to a comp as a team, for you can go as an individual and simply ask if someone is willing to help you out, when a word with the Contest Director should usually sort matters out. However, this does mean that you may well work with three different crews in one day, which may cause a little confusion, but then it's not such a bad way of meeting people. Comp fliers are generally speaking a pretty friendly lot, provided that you do not mind your leg being pulled. Always try

A well-thought-of 100S design is the *Algebra 2.5*, very much in line with current practice.

to be of assistance by timing and things should work out well, and what you learn by watching at the flight line will come in handy at a later date.

OPEN CLASS FLYING

The bulk of this chapter will deal with Open Class simply because it is the most popular class on the competition scene, and also because there is much that all the classes have in common when it comes to on-field organisation of people and

Developed from the 2m, the *Algebra 3m* is a good open model and a good choice for a third model to build.

Unusual low aspect-ratio 2 Metre model *Searcher* by Mark Kummerow.

tactics. The reasons for the popularity of the Open Class can be put down to four factors; firstly that the format of the competition is such that it can challenge the most competent and yet still be accessible to the newcomer, provided that he is willing to persevere. Secondly, its duration based task is fairly close to normal sport flying and therefore acceptable to a great many fliers. Thirdly that a good open model is a pleasure to fly just for sport purposes, and fourthly that the design freedom of this class is attractive to designers.

The task that Open sets you is to achieve the longest possible flight time within a ten minute time slot, with the addition of a landing within a 25m diameter circle for bonus points (50 for all

Nice looks and lightweight structure from Ross Garner of New Zealand and his *Time Machine*.

of the model in the circle & 25 for some part lying across the circumference). Scoring is by percentage slot. A minimum of three rounds are run and the top nine go forward to a two-slot fly-off with a time slot duration of fifteen minutes for each flight. The aim of the competition is to first get into the fly-off by virtue of being in the first nine and then try to beat the other qualifiers.

So what should you expect? If it's your first national comp, not a lot. The difference in overall ability between club and national comps can be intimidating if your club only has a few competitive fliers. Do not expect to set the world on fire, which you probably won't in your first season, unless you are exceptional. Content yourself with improving your scores and your consistency. Consistency is what really wins comps, not a range of scores that fluctuate wildly from the odd 1000 down to the low 200s. This consistent good flying is what marks the competitive flier out from less skilled fliers. If by the end of your first season you are regularly in the top third of the field then you will have done well. In your second season you should be knocking on the fly-off door and perhaps winning the occasional trophy.

At the field the procedure is that you first of all assemble your models and, time permitting, get a trim flight in. You should then hand your transmitter in to the Control Tent. The use of a Tx control tent virtually eliminates shooting-down incidents, provided that the Tx is switched off when it is handed in, which should be checked for on every occasion. You then wait and help until it is your turn to fly, but be aware of when it is so that you can get your flightline crew organised in time. It is very helpful if you and your crew are on the·same frequency (most clubs or groups usually have a preferred frequency), since you will then not have to fly against each other. You will often find that a group will be flying in consecutive slots, in which case it's best to arrange who is towing for who, spotting for who and so forth at the beginning of the round so as to avoid panic at the beginning of your flight preparation time.

When your slot is called you should ask your spotter or timer to collect your Tx whilst you yourself proceed to the flight line with your towman. Establish the wind direction and get the line laid out into wind, stretch and lift it to make sure that it hasn't become snagged in any plants (thistles are the worst) and then try to relax. When the Tx arrives switch on immediately and check the controls of the model, if satisfied check that the timekeeper has his two watches ready and that your spotter has the slot time watch ready. You now wait for the start of the slot.

Try to use the time to scan the sky for obvious signs of lift and get used to the air temperature so that you can spot any changes that might give a clue to conditions. Nerves can be quite a problem and sweating palms and knee trembling are quite common on the flight line, even among the experienced fliers. You will have to learn how to live with your nerves in your own way, which can be a valuable learning experience in itself. Some fliers find that they cannot fly to their best form unless the adrenalin is flowing, though generally its best to keep cool under pressure. What happens next depends upon many factors and comes under the heading of:

TACTICS & TECHNIQUES

Factor 1 – THE WEATHER. The great variable and cause of many frustrations. When the start horn sounds you should have a good idea of when you are going to start your flight. Much will depend upon your reading of the potential lift available. If you think that the air is good then there is no point in waiting and you may as well start immediately. It may well be that the other fliers in your slot feel the same way, in which case your choice is forced and you have to go. If, on the other hand, there is a cold wind blowing and you think that there is sink about it is best to wait until either the air warms up or the rest of the fliers decide to go. How you deal with the weather depends a great deal upon your experience and ability.

Factor 2 – THE MODEL. What model you are flying can determine to some extent your tactics. A 100S model in among the usual twelve footers has a lower potential duration if no lift is found and it is therefore no use going early in a dodgy slot. Wait until the rest of the field launch and then go. The longer they wait the less the potential duration will have to be and the less advantage to the big models.

On the other hand a big model (10ft. plus) is not limited so much by its potential duration and given the weather considerations mentioned above has much more freedom about when it can be launched.

A 2 Metre model in Open Class is almost certainly outclassed unless the lift is very good throughout all its slots (highly unlikely), or it is a super-lightweight in calm conditions. It is not worth flying a 2 Metre unless the wind is very high (25 mph plus) and even then it's likely to be outflown by a big heavyweight model.

Factor 3 – WINDSPEED. Great strides have been made in terms of structures and sections for flying in worse and worse conditions. If the wind is high and you have a model that you know from personal experience can take high windspeeds then proceed. But if you feel that either the model or your ability is not up to it then it's best to take the view that cowardice is the greater part of valour and scratch the flight. The repair time saved can be used to either build a suitable model or practise. At the other end of the Beaufort scale calm weather poses particular problems that are best solved by the use of a lightweight model with a fairly generous wing area. A full height launch can then be assured without necessarily giving your towman a hernia. Calm weather is a rarity on comp days.

Factor 4 – THE OPPOSITION. As a beginner you may well be somewhat overawed to be flying against well-known names. Don't be, as it's only by flying against them that you will learn. In some ways it's better, because they are unlikely to make fundamental errors of judgement and also they will be able to spot and use weaker lift that you may well not be aware of, so you can follow them and benefit from their greater experience. Be warned, though, some do not like to be followed and may well try to fool you into flying in a bad bit of sky! As a beginner you will not be known and watched so much, so you may well be able to get off early a few times and build up a lead, but if you do pick up a few scalps this advantage will not last long. Beginners are worrying to an experienced competitor because they are un-

The *Sunspot* was one of the large (around 10ft.) free-flight gliders so successful in the 1950 period. This example by the author makes an ideal, easy-to-fly trainer.

Another vintage favourite is the 1948 French design *Champ* **by E. Fillon. This one was built by Ron Hill and flown at Old Warden by Rod Holmes.**

predictable and can occasionally get away with blue murder. Beware of schoolboys and teenagers on the flight line – they can be embarrassing!

Factor 5 – MAXIMISING YOUR SCORE. Consistency is the real contest winner so you should try to score well on every flight. There are three basic times to launch: early, with the pack and late. Going early is like trying to bluff at poker, it often wins but can fail spectacularly if things do not work your way. There are three occasions when it can pay off. The first is when you think you have spotted lift that no one else has caught on to, when you obviously then have a time advantage. The second situation is when you have a poor score and decide to put the cat amongst the pigeons just for the devilment of discomfiting the rest of the field, and finally if you have an outside chance of making the fly-off but only if you can win the slot. In other words if you are going for broke.

The most consistent policy is to launch with the rest of the pack, hence the prevalence of the mass launches commonly seen. Going with the rest should ensure that you achieve a score in the 700 to 1000 range. Thus if you are sitting on a high two-round score you can relax a little and play the percentages game by going with the rest. As a beginner it's probably best to use this policy. Not many people delay their launch deliberately, since in good or average conditions it can be damaging pointswise. However, if you think that the rest are going up very early and in sink and if you trust your judgement, why not wait? It can happen that they are down in three minutes leaving you with a low flight time to beat and six minutes to do it in. You have the luxury of waiting for the sink to pass and if it doesn't you may be no worse off. Alternatively, you should be aware of the average flight times for the prevailing conditions and you may feel that a late launch may be a possibility.

Factor 6 – TIME PRESSURE. On a good lifty day it is quite possible to fly well beyond the end of the time slot by virtue of having gained a lot of height on a thermal. However, if this means that you have to lose a lot of height to stay within the slot time then you are in danger of either overflying your time or of destroying the model in your efforts to get it down. You should only gain enough height to be able to complete the slot, but allowance should be made for any possible sink. This means that you should be aware of how much time is remaining and adjust your height accordingly. Models have different height-losing capabilities. If you are up at 1000ft. plus shortly after the start of the flight, then it is worthwhile considering your plan of descent when there are some 4 minutes left and starting to come down to a manageable height shortly after, so that you have about 500ft. to lose in the last 2 minutes. This saves panic and makes the landing pattern easier to set up. The only way to learn how to handle time pressure is by experience and practice. Ask someone to count you down when you are practising your spot landings.

Factor 7 – LANDING POINTS. 100S rules do not require a landing in a set circle, a 75m radius landing area is used instead. The 25m diameter circle used by Open Class poses problems since failing to score the landing bonus on any flight can cost you the fly-off or the contest. It pays to practise your spot landings at the end of every flight, be it in sport or competition, and anyway it saves a lot of walking. Landing points assume different levels of importance depending upon the weather. If the flight times are short due to sink or high winds, then scoring the landing bonus becomes very important, since it may account for some 20 to 50% of your slot score. If the slots are being flown out to the full ten minutes then the landing bonus is less important, but desirable nonetheless.

There is one occasion when it may not matter if you miss the circle and that is if you have got away from the pack and managed to use a thermal that no one else has been able to get into. In these circumstances, so long as you outfly the rest by 50 seconds you are assured of 1000 points and you need not hit the circle, but then having got the upper hand you might as well grind the opposition down – they would try for a maximum score given the opportunity. One thing that needs to be emphasised is that it is pointless to perform a controlled crash into the circle just to ensure that you get the landing points. All that this achieves is the rapid deterioration of your model and the time-consuming task of keeping up with the running repairs that such arrivals cause. It is far better to go home with intact models and live to fight another day. Fly to survive. Airbrakes of some sort are virtually essential.

Factor 8 – FLY-OFF POINTS. The weather will determine the score that is required to qualify for the fly-off. A lifty day will mean that a three round score of 2900 points plus will be needed. A mixed day may bring the cut-off point down around 2800, and a poor or rough day may bring it down as low as 2600. Thus when considering your score and the task ahead of you it is worthwhile to consider the likely cut-off score, since it may help you to decide your tactics for the next flight.

Factor 9 – THE BARCS LEAGUE. Not an obvious factor but it can affect your tactics if you have a card in since you will then be trying hard on every flight in order to get a good score, regardless of the fly-off. The BARCS Leagues are regional leagues to decide the overall best fliers in the area. Doing well in them is prestigious enough in itself but if you want to take part in

Brian Johnson won the hand-launch R/C glider event at the '86 Scottish Nationals with *Mini-Whippet*, another RM plan.

Another H.L. glider, *Preacher Mite* by John Wesley. This type of flying is currently attracting increasing interest.

the big BARCS event, *Radioglide*, when it is in a popular region such as the Midlands then you will probably want to be assured of a place, and the only way to do that is to make the seeding based upon your league placing.

THE FLY-OFF

A separate heading because it really is a different ball-game. For one thing you are up against eight fliers who have proved their worth in the conditions, so the opposition is bound to be stiff. The pressure can be far greater, particularly if you want to do well. The best thing to do is to try your hardest and be consoled with the fact that the worst that you can do is to come ninth. The recent rule change lengthening the fly-off slots to 15 minutes has changed things a little, since thermalling ability becomes more important than in the standard slot. However, there is a little more leeway for waiting, if you dare. One thing that may be daunting is flying in a slot of nine flyers, when the only thing to do if your model is in the pack is to never look away from it and trust your spotter to tell you of any developments. This applies to any flight where you are flying in a group. When launching it may pay to wait until the traffic jam is clearing (or go first) if the towers are standing close together, which avoids a possible line crossing incident.

ODDS AND ENDS

The other competitions that may be of interest to the aspiring competitor are Hand Launch Gliders, 2 Metre and perhaps Vintage. Vintage at the time of writing is still a proposal for a fly-in type event but if it does work out it should provide some gentle low pressure competition for those interested in building such aeronautical antiques. Hand Launch is a side show at the moment and should develop nicely if all goes well. It may be that it would be better to concentrate on one class until that can be flown well before branching out. 2 Metre is in a parlous state although it may revive, but it may well be a little difficult for the raw beginner who may not be used to flying fast on a speed course. The class not mentioned so far is F3B, which is definitely an experts' class and should not be considered until a few years of Open and Slope flying have been put under your belt.

How to enter

You simply consult the contest diaries in the magazines and send off your details (name & frequency(s)) and enclose a stamped addressed envelope and the necessary money in some convenient form like a cheque. A full listing of competitions can be found in each March edition of the BARCS newsletter *Soarer*. Most competitors usually plan the year's campaign when this comes out. A useful idea is for a group of fliers who usually fly as a team to take it in turn each year to send off the whole group's entries as a block entry early in the season, which usually ensures that you get your desired frequency and helps the comp organisers to be able to plan with greater certainty.

So there you have it, competitive flying is a fascinating sport if you get involved in it. It must always be regarded as fun though, or it can be soul destroying. *Never ever take it too seriously*.

Probably the nicest thing about competition flying is the camaraderie amongst the fliers. It can be a really enjoyable day out to go to a comp; the fresh air (rain & wind), the healthy exercise (wheezing sessions after towing), the humour (leg pulling about your disasters), and the sheer joy of it all. What fun!

12 SLOPE SOARING COMPETITIONS

by Keith Thomas

You will rarely come across a thermal pilot who does not carry a stopwatch. In contrast, you will rarely come across a slope pilot who does. Once the beginner to thermal flying has mastered the basic skills of flying, it is natural that he wants to know how he is progressing, and the obvious method is to time each flight, perhaps note down the times, aim for his first six minute flight, then his first ten minutes, and so on. Soon he will be peering over his clubmate's shoulder, wondering what time he has to beat. The seeds of thermal competition are sown early. The beginner to slope flying, on the other hand, rarely has to concern himself with the length of a flight, as the lift is likely to be constant whenever weather conditions are stable. This is a fundamental difference between the two, apparently so intimately related, disciplines, and the different emphasis in each branch tends to attract different kinds of people. I know of a number of modellers who started out in thermal flying, mainly because there was a thriving thermal-based club nearby, who then gravitated to slope

Line-up of models at a recent slope meeting. How about some of the bigger models as pylon racers and aerobatic competitors?!

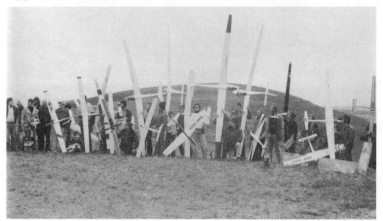

F

What sloping should be about on nice days. Max Fripp is about to launch Tony Baker's *Jantar 1*.

flying because they felt no inclination to compete. Equally there are fliers who move on from sloping to thermal work because they find the free-and-easy atmosphere of the typical slope session frustrating.

Elegant *Pik 20* by Roy Garner is a typical modern prototype kitted by Soaron Sailplanes and similar firms.

Hence the shortness of this chapter. The immense variety of slope activity, in terms of models, conditions and sites, means that the scope for competition is infinite. Nevertheless, slope contests are generally low-key affairs, with great emphasis on pleasure and little on pots. I'd hate to be responsible for changing that.

CLUB EVENTS

Most slope clubs run internal competitions, and that is the obvious place to start. Provided that your club is not dominated by a handful of win-at-all-costs merchants, the chances are that these events will be friendly, relaxed and not too intimidating even for the beginner. They can be loosely split up as follows:

Events requiring specialised model: aerobatics, pylon racing, scale, tailless.

Events open to any model: loops, spins, spot landings, elementary cross-country, multi-task.

Novelty: limbo, team racing, one-design, slalom.

The possible permutations are endless. Some clubs have developed a fixed annual programme of events, others leave the

Large scale slopers are becoming more common. This *Slingsby T46* **was designed and built by Chris Williams.**

A 1/5 scale *Slingsby Skylark* **from commercial plans by Norman Dean.**

make-up of the season's fun to the imagination of the competition secretary. If you should find that your club's calendar seems to be dominated by events requiring special models and high levels of skill, you might like to suggest watering them down with a few low-key events designed for more laughter and less furrowed brows. Here are a few guidelines for events which can be flown by any model.

"Maximum number of" competitions. The stopwatch is started at launch, and the pilot has, say, two minutes in which to complete as many recognisable loops, turns of spin, rolls, passes under an imaginary line, as possible. One judge operates the stopwatch, and decides whether each manoeuvre passes the test – you would be surprised what peculiar manoeuvres many pilots will insist is a loop. To avoid traffic jams, it is a good idea to place a further time limit of one minute in which the model has to land again within a specified (large) landing area. These ultra-simple competitions can be enormous fun, especially if conditions are not too good. The lift may be weak, and one loop too many may take you too low to land back on top inside the minute. Some models just refuse to loop, particularly on a competition day, and their potential for entertaining the troops is considerable.

The 4m span *Discus*, **also shown on page 178.**

Spot landing competitions are also deceptively simple. Add in a time factor – aiming to land on the spot exactly one minute after launch – and the event becomes both more difficult and more enjoyable. This is one of my personal favourites, as the best of pilots frequently boob, leaving the contest open to anybody. The spot can be just a bobble hat, or a carefully graduated series of boxes, which can be useful in persuading people to land in an orderly manner.

The drawback to most of these events is the poor use of air time. If conditions are good, it can be most frustrating to sit waiting for a stream of short flights before you have your own two minutes' worth. Two flight lines improves the situation at the expense of extra complication, but perhaps a better solution, when conditions are very good, is to try a different type of event, where several models can fly at once. A simple cross-country is the obvious candidate. This can be very enjoyable on a day with thermal activity as well as slope lift, and ensures long flight times for all. A course is laid out over the top of the hill, punctuated by turnpoints, and each pilot has to guide his model round the turns in sequence, and back to the finish. You can add a time element if you wish, or add specific flight manoeuvres at each turnpoint, but beware of adding complexity to the point where nobody wants to run the thing.

Slightly more complex, but within the scope of most clubs' organisational abilities, is the multi-task competition. This is a mixture of any two or more events which must be flown by the same model on the same day. Ideally the mixture exploits different aspects of the model's performance, such as aerobatics, cross-country, pylon racing and spot landing, but this places huge demands on the organiser. Maximum number of loops plus a timed pylon course plus a spot landing is much easier to arrange, and still pretty demanding. If you aim for a mixture which will favour the bog-standard sport model, you will not go far wrong.

NATIONAL COMPETITIONS

Once you have reached a reasonable level of proficiency at club level, you may well feel inclined to sample the rarefied air of national competitions. The relatively small number of nationally-advertised slope soaring events are centred around the following categories:

Cross-country
Cross-countries give everybody at least one long flight, barring

This scale vintage sailplane, the *L.O.100*, **is fully aerobatic. Designed by Cliff Charlesworth.**

accidents, and for this reason their popularity has remained high ever since their inception. With the large number of competitors the relative newcomer can "hide" comfortably in the pack and thus avoid the pressure of flying alone in front of judges. If the course is designed well, the beginner will get round most of it before its increasing difficulty forces him down. He can then sit back and watch how the experts do it. Highly recommended.

Precision aerobatics

Aerobatic events were once popular, but have almost disappeared now. As this is that rare event – a pure test of skill – it is inevitable that the same handful of names used to feature in the results list with monotonous regularity, but you can hardly blame the best pilots for being so good at it. I used to love going to watch the best pilots in the country doing their beautiful manoeuvres, so I hope the event will catch on again. The dearth of national events by no means mirrors the state of aerobatic flying at club level, as it is one of the most universally popular ways of spending time at the slope whenever conditions are favourable.

Multi-Task

More or less as described in the club events, but with everything shifted onto a higher level. There is generally a precision aerobatic section, so this is probably the best chance you will get to see good aerobatic pilots in action, albeit with models which are designed for compromise rather than specifically for aerobatics. The other sections are usually pylon racing and cross-country, with the result that the models are similar in most respects to competition F3B models. Most sections of these events are flown one-up, so you should be prepared to stand up in front of a judge, with many eyes squinting at your flying. Not an event at which a relative beginner is likely to acquit himself well.

Pylon racing

If you think your model is fast, take it along to a pylon race meeting. The heats are flown anything up to four-up, with many heats during the day whittling down the fliers to the fastest six or eight for semis and final. Very exciting to watch, very demanding to fly, and can be expensive in models, as mid-air collisions are not unknown. However, as so many models are flown at once, the beginner to competition flying can enjoy quite a lot of flying without feeling drowned in limelight. One word of warning – the top pylon fliers take things seriously, so don't get in their way.

Interesting twist with this scale *Scud 2* is that builder Martin Garnett is co-owner of the full-size machine!

Scale

Scale gliders are the source of my greatest pleasure, so you can expect a biased opinion here. If shiny pots are at the centre of many flying competitions, they have little place in scale events. The majority of scale modellers attend competitions simply to have a pleasant day out, fly their elegant models, and chew the fat with others of a like mind. This has led to the virtual disappearance of the old-style scale contest, with its two flights of two and five minutes, in favour of simple cross-country contests, perhaps with a gentle manoeuvre or two thrown in. With strict frequency control, several models can be flown at once, with the result that all entrants have loads of flying. I won't pretend that there is no hint of competition nerves, but most scale bods are all too keen to help, and the beginner will find a warm welcome. If you are at all interested in scale sailplanes, go along to one of the nationally advertised events and you will learn all you need to know about building and flying your first one.

APPENDICES

1   The Beaufort wind scale with recommendations for   170
    thermal soarers.

2   The F.A.I. limits and their application to current   172
    classes of models.

3   Current (1986) specifications for competition   173
    thermal soarers.

4   Possible training schedule/checklist for beginners.   174

5   Bungee recommendations.   175

6   B.A.R.C.S. Open Class Rules.   176

7   B.A.R.C.S. 100S Class Rules.   183

8   B.A.R.C.S. Thermal Soaring Achievement Programme.   187

9   B.A.R.C.S. Slope Soaring Achievement Programme.   189

    GLOSSARY OF USEFUL TERMS   192

    INDEX   204

APPENDIX 1

Beaufort Scale of Wind speed with Notes for Thermal Soarers

Beaufort No.	Terms used in Weather Bureau forecasts	Wind speed at 10 metres			
		in knots	in m/sec.	in mph	On land
0	Calm	Less than 1	Less than 0.3	0-1	Smoke rises vertically
1	Light	1-3	0.3-1.6	1-3	Smoke drifts
2	Light	4-6	1.6-3.3	4-7	Leaves on trees move slightly
3	Gentle	7-10	3.4-5.4	8-12	Flags flap, leaves and twigs on trees moving
4	Moderate	11-16	5.5-8.0	13-18	Dust raised, branches of trees swaying
5	Fresh	17-21	8.1-10.7	19-24	Small trees swaying
6	Strong	22-27	10.8-13.8	25-31	Large branches swaying
7	Strong	28-33	13.9-17.1	32-38	Large trees swaying, walking against wind is difficult
8	Gale	34-40	17.2-20.7	39-46	Twigs and small branches breaking, leaves stripping, dust storms
9	Gale	41-47	20.8-24.4	47-54	Large branches breaking, roof damage likely, hard to stand
10	Whole gale	48-55	24.5-28.3	55-63	Whole trees uprooted, some damage to many buildings
11	Whole gale	56-65	28.4-33.5	64-72	Widespread damage to buildings and vegetation
12	Hurricane Cyclone	Above 65	Above 33.6	73-82	Widespread disaster, houses destroyed, forests uprooted, etc.

Suitable Models	Useful Span	Loadings	Launch Method	Comments
Lightweight floaters medium weight models	6-14ft	5-9ozs/ sq.ft	Towline or winch required. Bungee gives half height launch	Too calm for comfort
Medium weight models	6-14ft	7-10ozs/ sq.ft	Towline or winch. Bungee good for light-weight models	Ideal cond-itions if sunny
Medium weight models	6-14ft	8-10ozs/ sq.ft	Towline or winch. Bungee good for up to medium weight models	Good cond-itions
Ballasted medium weight models and heavyweights	6-14ft	9-12ozs/ sq.ft	Towline. Bungee will be stretched at top of launch	Reasonable if sunny
Strong heavy-weight models with heavy ballast	6-8ft	14-18ozs/ sq.ft	Towline. Anything will tow up	Bad, fly only if necessary
Very heavy models	6-7ft	18ozs/ sq.ft +	Towline with heavy line	Awful, retire to nearest hostelry or workshop

— *Stay at home and batten the hatches!*

N.B.: Add 50% onto any ground level wind meter reading to account for wind shear, less if surrounding countryside is flat and unobstructed.
Lightweight, medium weight and heavyweight refer to the airframe structural strength not wing loading.

APPENDIX 2

F.A.I. Limits to Model Size

Max. Flying Weight	5 Kilograms	(11 lbs)
Max. Flying Surface Area	150 sq decimetres	(16.146 sq ft)
Max. Flying Surface Loading	75 gms/sq decimetre	(24.5 ozs/sq ft)
Min. Flying Surface Loading	12 gms/sq decimetre	(4 ozs/sq ft)

Table of Models for Various Competitions

class	span	wing area	controls	launching	suitability for beginners
OPEN	unlimited	FAI limit	any	towline	reasonable
100S	100 ins.	800sq.ins	rud/elev airbrakes	towline	good
2Metre	2 metres	FAI limit	rud/elev.	towline	fair
F3B	unlimited	FAI limit	any	power winch or towline	poor
Hand Launch	unlimited	FAI limit	any	thrown by hand	good
Vintage	as per plan*	as per plan*	any	towline	good

*Model may be increased in overall size

APPENDIX 3

Table of Current Model measurements for various Classes (1986)

Class	span	wing area	controls	loadings with ballast	common dry loading
OPEN	10-15ft	7-12 sq.ft	rud/elev/ brakes (flaps/ail.)*	6-16 ozs/ sq.ft	8 ozs/sq.ft
100S	100 ins.	5.55 sq.ft	rud/elev airbrakes	6-16 ozs/ sq.ft	8 ozs/sq.ft
2Metre	2 metres	4.5 sq.ft	rud/elev	6-16 ozs/ sq.ft	9 ozs/sq.ft
F3B	110-130 ins.	6-9 sq.ft	rud/elev/ brakes/ flaps/ail	9-20 ozs/ sq.ft	12 ozs/ sq.ft
Hand Launch	50-65 ins	2-3.5 sq.ft	rud/elev (flaps)*	not appli-cable	4-7 ozs sq.ft
Vintage	50-140 ins	2-12 sq.ft	rud/elev	not appli-cable	4-10 ozs/ sq.ft

*Not too common but used sometimes

APPENDIX 4

Proposed Training Schedule for Thermal Soaring

Instructors
Initials

1. Straight and level Flight – without undue loss of height

2. Stall Recovery – without undue loss of height

3. Right Hand Turns (Wide) – without undue loss of height

4. Left Hand Turns (wide) – without undue loss of height

5. Figures of 8 Crosswind (wide) – without undue loss of height

6. Tight Turns Right & Left – without undue loss of height

7. Thermal Turns Right and Left – without undue loss of height

8. Downwind Flying – without stalling

9. Flying Towards Flier – without loss of orientation

10. Approach For Landing – judgement and placing

11. Landing – tidily and without incident

12. Bungee and Towline Technique

13. Loops

14. Stall Turns

15. Spin Recovery

Each section to be initiated when Instructor is satisfied that the Trainee is competent in particular section.

APPENDIX 5

Instructions for 5/16" & 3/8" Heavy Duty Hi-Start/Bungee *

Length of Amber tubing approx 100 feet.

Amber tubing will stretch up to six times its unstretched length, ordinary rubber will normally only stretch three times, this is where Amber tubing has the advantage. The ability to exert a long steady pull will gain much more height, than the quick short burst of power from the ordinary rubber bungee. Any Hi-Start/Bungee will give best results when there is a breeze blowing, when releasing the model always throw hard in the direction of pull, extra height can be gained and there is less chance of a bad launch.

To make up into a complete launching system, first drill a hole within ¼" of one end of the wooden dowel supplied, then insert dowel as far as possible into surgical tubing. Make up a loop of 50lbs nylon line, pass this through hole in dowel, this loop is then anchored to the ground by a stake. Then make up a line approx 120m in length of 50lbs nylon line, attach this to the other dowel. For best results in operation, some breeze should be blowing. The ⅜" extra heavy duty tubing is intended for large thermal soarers with wing spans up to 14 feet and a weight of 6lbs or the extra heavily loaded F3B type of model with strong wings. Do not use on a model below 100" span with lightly constructed wings.

Suggested amount of stretch (in normal paces). This is a rough guide for the inexperienced. Factors such as wing span, wing area, wing loading and structural strength will all affect the pull required. If in doubt always err on the high side, understretching will usually cause more problems than overstretching.

5/16" Tubing

Model Weight	Paces Stretch
1.5 to 1.75 lbs	30 to 40
1.80 to 2.25 lbs	50 to 65
2.30 to 2.75 lbs	70 to 80
2.80 to 4.00 lbs	85 to 95
4.25 to 5.00 lbs	95 to 110
5.00 lbs plus	120 maximum recommended pull

Large heavier models will not gain full height unless there is a breeze of approximately 6 mph or more.

3/8" Tubing

Model Weight	Paces Stretch
2.75 to 3.25 lbs (normal type wings)	60 to 70
2.75 to 6.00 lbs (F3B type strong wings)	100 to 120
3.30 to 4.00 lbs (normal type wings)	75 to 90
4.25 to 5.00 lbs (normal type wings)	90 to 100
5.25 to 6.00 lbs (normal type wings)	100 to 120

Care and Storage

Try to avoid dragging the Hi-Start tubing over rough or sharp objects. Should you break the tubing it is readily joined by inserting a wooden dowel, this will have no detrimental affect on its performance. Storage, always keep in the dark, out of sunlight, preferably in the cool.

Information courtesy of Edmonds Model Products.

APPENDIX 6

Section 2.3

B.A.R.C.S. Rules for Open Class Radio-controlled Thermal Soaring Competition 1985 Edition
Percentage Slot Scoring System

1. **Object**
    To provide standards for the competition of radio-controlled soaring gliders.

2. **Flying Site**
    The competition must be held on a site having reasonably level terrain which will minimise the possibility of slope and wave soaring.

3. **B.A.R.C.S. League**
    A competition will be granted league status in advance at the discretion of the appropriate B.A.R.C.S. Area Representative who is empowered to endorse minimal deviations from only those rules marked with an *, to suit the prevailing circumstances. Only contests publicised in "Soarer" or the National Model press before the date will be granted league status.

4. **Model Characteristics**
    a. Maximum surface area: 150 sq decimetres (2325 sq inches)
        Maximum flying weight: 5 kilograms (11.023 lbs)
        Maximum surface loading: 75 grams/sq decimetre (24.51 ozs/sq.ft)
        Minimum surface loading: 12 grams/sq decimetre (3.95 ozs/sq.ft)
    b. The nose radius of the model shall not be less than 5mm (3/16th in.) radius measured tangentially at all intersecting surfaces.
    c. Radios shall be able to operate simultaneously with other systems on the same frequency band. The preferred separation capability is 20KHz for 27MHz systems, 10KHz for 35MHz systems and 25KHz for UHF systems. The minimum separation capability is 50KHz for 27MHz and UHF systems and 20KHz for 35MHz systems.
    d. Any device for the transmission of information from the model to the pilot by means of radio equipment is prohibited.
    e. A competitor may use a maximum of TWO models.
    f. Component parts of the two models may be interchanged but not with those of other competitors.
    g. All ballast must be carried internally and fastened securely within the airframe.
    h. Any competitor operating equipment transmitting outside the U.K. licensed frequency bands at a BARCS League event shall be disqualified from the league for that year.

5. **Ownership of Models**
   a. Any individual model may only be flown by one entrant in any particular competition.
   b. The entrant shall be the genuine owner of the model and as proof of ownership the entrant's name or B.A.R.C.S. number shall be displayed on the wing of the model in a permanent and prominent manner.

6. **Protests**
   a. Any protests are to be made in the first instance to the Contest Director accompanied by a protest fee equal to the entry fee as soon as possible.
   b. If the protestor is not reasonably satisfied after the initial protest, a further written protest may be lodged with the appropriate B.A.R.C.S. area representative (or his appointed deputy). The consequences of this further protest shall be limited to changing the allocation of B.A.R.C.S. league points.
   c. Protests against an area representative's decision shall be addressed in writing to the B.A.R.C.S. Committee and sent to the Secretary.

7. **Competitor and Helpers**
   a. Each competitor is permitted TWO helpers.

8. **Competition Flights**
   a. The competitor will be allowed at least THREE official flights.
   b. The competitor will be allowed TWO attempts at each official flight.
   c. There is an official attempt at a flight when the model has left the hands of the competitor or his/her helper under the pull of the launching apparatus.
   d. If for any reason the official flight is timed at less than 60 seconds in duration the competitor will be allowed the second attempt which must be made immediately and within the allocated time slot.
   e. All flights to be timed by two stopwatches and in the event of both stopwatches malfunctioning the flight will count as zero.

9. **Cancellation of a flight and/or disqualification**
   a. The flight is cancelled and recorded as a zero score if the competitor used a model not conforming to any items of Rule Number 4. In the event of intentional or flagrant violation of the rules in the judgement of the Contest Director, the competitor may be disqualified.
   b. The flight is cancelled and recorded as a zero score if the model loses any part in flight, except where this occurs as the result of a mid-air collision with another model or towline.
   c. The loss of any part of the model during the landing (touchdown) will not be recorded.
   d. The flight is cancelled and recorded as a zero score if the model is piloted by anyone other than the competitor.
   *e. The flight is cancelled and recorded as zero score if some part of the model does not land within 75m of the centre of the designated landing circle.

10. **Organisation of the Flying Slot**
    a. The flying order for the initial three rounds shall be arranged in accordance with the radio frequencies in use to permit as many simultaneous flights as possible.
    b. The flying order must be scheduled in Rounds sub-divided into Time Slots.

G

Unusual planform on this *Discus* replica owned by Bob Page.

A design available in two sizes is *Hi-Fibre*. Here the designer, Dave Kite, holds the 16ft. (nearly 5m) variant.

c. The flying order shall be determined by a Matrix system (See note at the end of these rules regarding availability of matrix sets) that minimises situations where competitors fly together more than once.

d. Entry on the day of the contest will only be accepted if a vacant position is available in the Matrix.

e. A competition number, derived from the Matrix, must be allocated to each competitor which must be retained throughout the first three rounds.

f. Competitors are entitled to a minimum of five minutes preparation time which is counted from the moment he /she is called to take position at the designated launching area.

g. The organisers must indicate very positively the start of Slot Time both audibly and visually, see Appendix for details.

h. The Slot Time shall be exactly 10 (ten) minutes duration.

*i. Audible and visual signals must be given when 8 (eight) minutes of the Slot Time has elapsed.

*j The end of the Slot Time must be very positively indicated both audibly and visually, as for the start.

k. Any model airborne at the completion of the Slot Time must land immediately.

11. **Control of Transmitters**

a. The Contest Director will not start the competition flying until all competitors have handed over ALL transmitters to the organisers.

b. Failure to hand in a transmitter before the official starting time of the contest may result in the competitor forfeiting his/her first round flight.

c. Any test transmission during the contest without permission of the Contest Director is forbidden and could result in disqualification.

d. The competitor must hand over his/her transmitter to the designated official (usually the timekeeper) immediately after finishing the flight.

12. **Launching**

a. The launch of the model may be by:
   i. Hand held towline.
   ii. Winch devices (mechanically or hand powered).

b. Towlines or power winches for each flyer must only be run out during the competitor's five minute preparation period and must be retrieved by the end of the slot.

c. The use of a bungee (Hi-start) consisting of an elastic member is not permitted.

d. Towing by moving vehicles such as bicycle, car, motorcycle and radio-controlled powered aircraft, is not permitted.

e. The length of the towline for hand held towing shall not exceed 150m when tested under a tension of 2kg (4.4lbs).

f. The effective length for launching by winch devices shall not exceed 150m when tested under a tension of 2kg (4.4lbs).

g. i. To facilitate observation and timing the towline must be equipped with a pennant having a minimum area of 5 sq decimetres (77 sq. ins.).
   ii. A parachute may be substituted for the pennant providing it is not attached to the model and remains inoperative until the release of the towline.

h. Tow-men must remain within the designated towing area.

    i. No tow-person shall cast off a model by means of the release of the towline by the tow-person. Penalties will be at the Contest Director's discretion and may be imposed on the pilot and/or the tow-person.

    j. Any model launched prior to the start of the Slot Time must be landed and relaunched within the Slot Time. Failure to comply will result in cancellation of the competitor's score for that Round.

13   **Landing**

    a. The landing targets shall comprise a minimum of TWO circles each of 25m diameter. Organisers must allocate particular frequencies (colours) to each circle.

*b. The circles must be laid out in a line at right angles to the wind direction.

    c. Competitors and officials (timekeepers) must remain on the upwind side of the landing circles during the landing process.

    d. Competitors may only retrieve their models on completion of the landing providing they do not impede other competitors and models.

14.   **Scoring**

    a. The flight will be timed from the moment of release from the launching device to:
        i. The moment the model first touches the ground.
        ii. The moment the model first touches any object in contact with the ground. Parts of launching devices extending away from the ground shall not be interpreted as objects in contact with the ground.
        iii. Completion of the slot time.

    b. The flight score will be composed of ONE point for each FULL second of flight time.

    c. A penalty of 30 points will be deducted from the flight score for overflying the end of the Slot Time for up to a maximum of ONE minute (sixty seconds).

    d. A zero score will be recorded for overflying the end of the one minute penalty time.

    e. A bonus of fifty points will be awarded, providing ALL parts of the model come to rest completely within the landing circle.

    f. A bonus of twenty five points will be awarded, providing some part of the model which has not become detached during the landing, comes to rest within the landing circle.

    g. No landing bonus points will be awarded if the model overflies the end of the Slot Time.

    h. The competitor who achieves the highest aggregate of points, i.e. flight points plus landing bonus points/less penalty points, will be awarded a corrected score of one thousand points for that Slot.

    i. The remaining competitors in that Slot will be awarded a percentage of the slot winner's total score calculated from their own total score.

15.   **Final Classification**

    a. At the completion of all rounds the 9 (nine) competitors with the highest total of percentaged scores must perform in a Fly-off to produce the Final Competition placings by one of the following methods:
        i. Three further six-competitor slots whereby all nine finalists will compete against each other at least once. In the event of a fixed frequency clash in qualifying for the fly-off the competitor with the

lowest total score unable to change frequency must drop out in favour of the next competitor.

    ii. Two further slots whereby all nine finalists compete simultaneously against each other twice.

  b. The Flight and Scoring Rules for the Final Fly-off differ from the initial rounds in no way other than the slot time being increased to 15 minutes and an audible warning being given at 13 minutes.

Note:

The appendix which follows gives some general advice to contest organisers.

Very detailed notes, together with sets of matrices covering 20 to 110 competitors are available on request from the Newsletter Editor, and anyone organising a percentage slot contest for the first time is strongly advised to read these.

Send S A E., stamped to allow for weight of 160 grams (4th rate of second class postage) to Newsletter Editor, B.A.R.C.S., requesting "Notes for Organisers".

Appendix

Advisory information to help those organising competitions to these rules.

1.   **Slots**

The organisers must ensure that each competitor has no doubt about the precise second that the Slot Time starts and finishes.

Visual indication may be by the raising of a flag or coloured board situated near the Contest Control.

Audible indication may be by motor horn, aerosol horn or bell etc. It should be remembered that sound does not travel far against the wind therefore the positioning of a noise source must be given some thought.

To be a fair competition the *minimum* number of fliers in any one slot is 4 (four). As the competition proceeds, some competitors will be obliged to drop out, for some reason. When a slot comes up with only 3 contestants in it the organisers must move up a flier from the later slot ensuring he has not flown against any of the three in previous rounds and of course that his frequency is compatible.

2.   **B.A.R.C.S. League**

  a. Points for the League are awarded depending on the final placings at the end of Round Three.

  b. It is the responsibility of the competitor to ensure his/her League Score Card is correctly filled in and handed to the organisers before the start of the contest.

  c. The Contest Director is requested to enter the results on each Card after the contest and pass them on to the B.A.R.C.S. League Co-ordinator not more than two weeks after the event. Only results returned on standard league score cards will be accepted. Postage costs for sending League cards will be refunded by BARCS upon request.

  d. In the event that the competition is terminated short of three rounds, due to bad weather or insufficient time for example, league cards must be returned to their owners. The competition would be declared void in respect of the league.

3.   **The B.A.R.C.S. Matrices**

  a. A Matrix must be employed to minimise situations where any competitor flies against another more than once ,except in the fly off. (It is recognised that, in practice, with certain "awkward" numbers of competitors, or where more than three rounds are flown, a degree of "clashing" will be unavoidable , but this should be kept to the minimum).

b. The method by which each competitor is given his/her competition number from the Matrix is left to the organisers.

c. Once the contest has started neither the Matrix Table nor competition numbers must be changed.

d. In order to minimise the time needed to run the contest, it is very important that the matrix which gives the minimum number of slots per round, with the maximum possible competitors in each, should be selected, and the number of frequency groups adjusted accordingly.

4. **The Fly-Off**

As Rule 15 states the nine competitors in the Fly-Off must fly against each other to reduce the luck element.

The simplest method for the Fly-Off is to fly all nine competitors simultaneously, twice, so that they can all perform in the same block of air. Whenever split frequencies are being used, whether in the main contest or the Fly-Off, all pilots should be confined to the designated area when performing and not be allowed to move around the field. This will reduce the chance of adjacent channel interference. If a frequency or manpower problem prevents a simultaneous fly-off, three slots of six competitors each should be flown so that each competitor meets every other at least once.

5. **Time-Keeper Duties**

Organisers should make sure all who are to act as time-keepers are fully aware of just how important their duties are and to make certain they are conversant with the rules particularly those that require quick positive action in order not to jeopardize a competitor's chances in the competition.

The time-keepers will be responsible for handing transmitters to competitors prior to the start of the slot time and for returning them to control at the end of the slot or flight, whichever is the earlier.

The organisers must ensure that an official is nominated to note any competitor who overflies the end of the slot time and to time the excess time.

6. **Safety**

Finally, but not least, previous experience has made everyone very conscious of taking the utmost care to effect safety towards competitors, including helpers, officials, and not forgetting spectators. Two important areas to consider are the careful layout of the flying field and controlling ie the number of persons on the field at all times.

Remember the appropriate B.A.R.C.S. Area Representative will answer any queries on the Rules and Appendix.

APPENDIX 7

Section 2.4

Rules for 100S (Standard) Class R/C Gliders 1985 Edition

1. **Objective**
   Competition of standardised radio controlled gliders.

2. **Model Characteristics**
   a. Maximum span of wing 100″.
   b. Maximum total projected area of wing 800 sq.ins.
   * c. The use of flaps, airbrakes or other moving lift control devices is prohibited.
   * d. The maximum number of operation servos shall be two.
   e. Radios shall be able to operate simultaneously with other systems on the same frequency band. The preferred separation capability is 20KHz for 27MHz systems, 10KHz for 35MHz systems and 25KHz for UHF systems. The minimum separation capability is 50KHz for 27MHz and UHF systems and 20MHz systems.
   f. The competitor may use no more than two gliders during the contest.
   g. The nose radius of the model shall be not less than 5mm (3/16th in) radius measured tangentially at all intersecting surfaces.

3. **Certification**
   a. It is the competitor's responsibility to ensure that the glider complies with the class specification.
   b. The competitor, when required, shall produce a certificate signed by himself and countersigned by a second party, stating the measured wing dimensions and areas of his glider. Failure to produce a certificate, or producing a faulty certificate, may lead to disqualification.
   c. At the Contest Director's discretion, or upon demand of a competitor, any model may be checked to ascertain that it complies with the specification.

4. **Control of Transmitters**
   a. The organiser cannot begin the competition flights until all competitors have handed their transmitters over to the organisers. Failure to hand in a transmitter before the official starting time of the contest may result in disqualification.
   b. Any test transmission during the course of the competition without the permission of the organisers is forbidden, and entails disqualification.
   c. The competitor must hand over his transmitter to the designated official immediately after finishing his flight.

5. **Organisation of Flights**
   a. The flying order for the first four rounds shall be arranged in accordance with the radio frequencies in use to permit as many simultaneous flights as possible.
   b. The flying order must be scheduled in Rounds sub-divided into time slots.
   c. The flying shall be determined by a matrix system that minimises situations where competitors fly together more than once.
   d. Entry on the day of the contest will only be accepted if there is a vacant position available in the matrix.
   e. A competition number derived from the matrix must be allocated to each competitor and this must be retained throughout the first four (4) rounds.
   f. Competitors are entitled to a minimum of five minutes preparation time which is counted from the moment he/she is called upon to take up position in the designated launching area.
   g. The organisers must indicate very positively the start of the slot time both audibly and visually.
   h. The slot time shall be of exactly eight (8) minutes duration.
   i. Audible and visual signals must be given when six (6) minutes of slot time have elapsed.
   j. The end of the slot time must be very positively indicated both audibly and visually as for the start.
   k. Any model airborne at the completion of the slot time must land immediately.

6. **Competition Flights**
   a. The competitor has the right to four official flights.
   b. There is an official attempt at a flight when the model has left the hands of the competitor or his assistant under the pull of the launching apparatus.
   c. The competitor has the right to two attempts at each official flight providing that the first attempt is less than 60 seconds duration.
   d. In every instance of the flight not being judged by fault of the official timekeeper, that attempt it void and the competitor is allowed another attempt.
   e. Repeat attempts shall be taken immediately, or at a time designated by the Contest Director.
   f. Each offical flight score shall be determined by the competitor's last attempt at that official flight.

7. **Cancellation of Flight or Disqualification**
   a. The flight is cancelled and recorded as a zero score if the competitor used a model not conforming to rule 2. In the case of intentional or flagrant violation of the rules in the judgement of the Contest Director, the competitor(s) may be disqualified.
   b. The flight is cancelled and recorded as a zero score if the model loses any part in flight. The losing of a part during landing or in accidental collision in flight with another model or towline is not taken into account.
   c. The flight is cancelled and recorded as a zero score if flown by anyone other than the entrant.
   d. The flight is cancelled and recorded as a zero score if the model does not land within 75m of the landing target.

**Equipment installation in two slim-fuselaged high-performance soarers,
left on** *Algebra 3M* **and right on F3B model by Dwight Holley.**

8. **Launching**
   a. The launching may be by hand tow only. Towing by moving vehicles, such as bicycles or automobiles is not permitted.
   b. The length of the hand towline shall not exceed 150m when tested under tension of 2kg (4.4lbs).
   c. To facilitate observation timing, the hand towline must be equipped with a pennant or parachute having a minimum area of 5 sq. decimetres.

9. **Scoring**
   a. The flight will be timed from the moment of release from the launching device to:
      i. The moment that the model first touches the ground.
      ii. The moment that the model first touches an object in contact with the ground. Parts of launching apparatus extending away from the ground shall not be interpreted as objects in contact with the ground.
      iii. Completion of the slot time.
   b. The flight score will be composed of ONE (1) point for each FULL second of the flight time.
   c. A penalty of eighty (80) points shall be deducted from the flight score for overflying the end of the slot time by up to a maximum of 1 minute (sixty seconds).
   d. A zero score will be awarded for overflying the end of the one minute penalty time.
   e. The competitor who achieves the highest aggregate of points – i.e. flight points less penalty points, in each slot, will be awarded a corrected score of one thousand (1000) points for that round.
   f. The remaining competitors in each slot will be awarded a percentage of the slot winner's score calculated from their own points divided by the slot winner's points and multiplied by 1000.

10. **Final Classification**
    a. At the completion of ALL rounds the 9 (nine) competitors with the highest total of percentaged scores must perform in a Fly-off to produce the final competion placings by one of the following methods:
       i. Three further six competitor slots whereby all nine finalists will compete against each other at least once.
          In the event of a fixed frequency clash in qualifying for the Fly-off, the competitor with the lowest total score unable to change frequency must drop out in favour of the next competitor.
       ii. Two further slots whereby all nine finalists compete simultaneously against each other twice.
    b. The Flight and Scoring Rules for the final Fly-off are exactly as for the first four Rounds.

11. **Site**
    The competition shall be flown on a site having reasonably level terrain that does not include the possibility of wave or slope soaring.

   * A recent change of rules has altered 2.c and 2.d. The model may now have airbrakes fitted and can have up to three servos installed.

APPENDIX 8

Section 3.2

Thermal Soaring Achievement Programme

1. **Aim of the Scheme**
   The award scheme is designed to appeal to the soarer who does not wish to travel to contests, but at the same time would like to progressively improve his flying abilities and furnish concrete evidence of this improvement.

2. **Awards**
   Attainment of the requirements for the various levels of the programme is recognised by the award of a certificate and the appropriate level stickers for display on the member's models (extra stickers may be purchased by those qualified for that particular grade only), also by the award of small trophies for the diamond and double diamond grades.

3. **General Requirements**
   a. The model must conform to the general maximum weight and size limits as laid down in FAI rules.
   b. The flights must take place on *level ground* at least two miles from obvious slope lift.
   c. Launch must be by 150m hand towline or 150m bungee stretched to 2kg.
   d. Flights are not precision time, they may exceed the required flight time by any margin.
   e. Claims must be made sequentially – i.e. bronze tasks must be completed before a claim may be entered for the silver, and so on.
   f. All flights must be witnessed, note the special requirements for the double diamond.

4. **Tasks**
   a. **Bronze Level**
      i. Two soaring flights of 5 minutes minimum duration.
      ii. Five landings within 10 paces of a nominated target.
   b. **Silver Level**
      i. Two soaring flights of 10 minutes duration.
      ii. Five *consecutive* landings within 5 paces of a nominated target.
   c. **Gold Level**
      i. One soaring flight of 10 minutes minimum duration.
      ii. One soaring flight of 20 minutes minimum duration.
      iii. One cross-country flight of one mile, *or* a second soaring flight of 20 minutes minimum duration.
   d. **Diamond Level**
      i. One soaring flight of 40 minutes minimum duration.

ii. One goal and return cross country flight of two miles (i.e., one mile out, one mile back).
iii. One flight around a triangular course with sides of 300 metres (total 900 metres).

e. **Double Diamond Level**
   i. One soaring flight of 3 hours minimum duration.
   ii. One goal and return flight of 20 kilometres (i.e., 10 kilometres out, 10 kilometres back).
   iii. One cross country flight of 25 kilometres minimum.
   iv. One flight over a triangular course having two sides of 5 kilometres each and the third side at least 28% of the total distance around the course.
   v. an article for the Newsletter describing the flights must be provided.

Claims

All Double Diamond claims must be witnessed by two B.A.R.C.S. members of at least one year's standing.

Claims at Bronze to Diamond level should be made when all tasks for the grade have been completed. For the Double Diamond, each task should be claimed separately as it is completed.

Levels may only be attempted in order – i.e. to claim the silver, you must already hold the bronze and so on.

All enquiries and claims regarding the Thermal Achievement Programme should be sent to the Records and Trophies Officer, B.A.R.C.S. ALL correspondence requiring a reply should enclose a stamped addressed envelope. Where an achievement certificate is to be returned the envelope should be at least 9" × 6".

APPENDIX 9

Section 3.3

Slope Soaring Achievement Programme

The B.A.R.C.S. Slope Soaring Achievement programme has been drawn up to provide a series of tasks of a varied nature and, while progressing through the schedule, increasing complexity. Many of these tasks have been taken from well proven programmes used throughout the world.

When following the programme, a slope soaring pilot will become increasingly proficient and, on attaining the Diamond Hawk award may be able to regard himself as a highly competent slope pilot.

General Requirements

1. The tasks shall be witnessed and judged by two persons who are conversant with slope soaring aerobatic manoeuvres by virtue of experience at flying in competitions.

2. The flier or helper will hand launch the model. The flier will perform the prescribed manoeuvres for the grade being attempted in the order set out in the schedule.

3. Each judge will give a score of up to 10 points to each manoeuvre as it is performed. The average of these points will be the score awarded for that manoeuvre. Each manoeuvre will qualify for acceptance if it is awarded a minimum of 5 points out of 10.

4. It should be noted that the quality of the execution of the manoeuvre is the basis of scoring, i.e. if in Silver 1(d) only a 5 second inverted flight is achieved and all other aspects are good, then a qualifying score of 5 points will be awarded. If the 5 second flight was part of an otherwise unsatisfactory manoeuvre, then this will not qualify.

5. Should a flier fail to perform any manoeuvre to the required standard, he shall be immediately advised of the fact by the judges. To qualify for the grade being attempted, the flier will have to restart the whole sequence again from the launch of the model.

6. The judges' decision as to the satisfactory performance of any manoeuvre shall be taken as final. No protests are allowed or are to be entered into.

7. The full requirements for any grade must be completed before any attempts can be recorded in the next grade.

8. All applications to the B.A.R.C.S. Slope Representative for recognition of success in the grades shall be made in writing and must be verified by the officiating judges.

PROGRAMME

Bronze Hawk Grade

1. A slope soaring flight the duration of which shall be 15 minutes minimum.

During the flight the following manoeuvres shall be performed:

a. One large circle (right or left) horizontal.

b. Two consecutive loops across wind.

c. 10 seconds straight and level across wind.

d. Landing pattern (rectangular).

Silver Hawk Grade

1. A slope soaring flight as per Bronze Hawk to include the following manoeuvres:

   a. One horizontal 8.

   b. Three consecutive loops across wind.

   c. One stall turn (right or left).

   d. 10 seconds' inverted flight, across wind.

   e. One axial roll across wind, (right or left).

2. 3 spot landings within 5 paces of a predetermined spot. (Measured when model comes to rest).

3. A goal and return flight of ½ mile (800m) total distance, predetermined prior to start.

Gold Hawk Grade

1. A slope soaring flight of 1 hour minimum duration to include the following manoeuvres:

   a. 3 consecutive axial rolls across wind.

   b. One double Immelmann.

   c. 3 consecutive outside loops (bunts).

   d. 3 turn spin.

   e. One vertical 8.

   f. One Cuban 8.

   g. Inverted flight for 2 minutes duration.

   h. Landing pattern (rectangular).

2. 3 spot landings within 2 paces of a predetermined spot.

3. A goal and return flight of 1 mile (1.6km) total distance, predetermined prior to start.

4. A slope duration flight of 2 hours.

5. A short description of the model or models used and an account of your attempts at various tasks for this grade.

Diamond Hawk Grade

1. A slope soaring flight of 1 hour minimum duration to include the following manoeuvres:

   a. Slow axial roll to left immediately followed by slow axial roll to right.

   b. Six turn spin.

   c. Four point axial roll.

   d. Figure M.

   e. Horizontal 8.

   f. Vertical 8.

   g. 3 minutes inverted flight.

   h. 6 consecutive loops across wind.

   i. One double Immelmann.

   j. 3 consecutive axial rolls.

   k. Landing pattern (rectangular).

2.   A goal and return flight of 2.5 miles (4 km) total distance, predetermined prior to start.

3.   A short account of your attempts at the tasks contained in the Diamond Hawk award.

How to take part
Write to the B.A.R.C.S. Slope Representative, ENCLOSING A STAMPED ADDRESSED ENVELOPE. He will send you full details of the programme and notes to assist judges in marking your attempts. At each level you receive the appropriate award to display on your model.

First BARCS Slope Gold . . .

Extracts from the report by Chas Gardiner on his attempts.

'The first task attempted was the "goal and return", on which I was joined by the newsletter Editor flying his "Solitaire". My model was an O/D "Fledermog", 120 inch span, conventional polyhedral built-up wing, sheet covered, with balsa fuselage and A.M.T. Loading 10 ounces per square foot.

The job was made more difficult in that the site used was Mam Tor, near Castleton in Derbyshire. Imagine an "L" shape with us launching into wind at the bottom right of the base. Head left to the "heel" and forward into wind up the vertical which is in fact the spine of the ridge which curves for a couple of miles right and left between Edale and the Hope Valley. Gaining height from the start was easy but pushing forward from one pocket of lift to the next was exceedingly tricky. At one stage both models were pinned down on *opposite* sides of the ridge for a long time until thermal lift came through. This epic flight took place in fact about 18 months ago after completion of the "silver".

Spot landings were the speciality of the 36 inch "Obelix" aileron/elevator model. The rules don't specify that they should be consecutive or the quality, just the distance from the spot. This is a pity as the temptation is to "dump" the model, which is not to be encouraged. Still . . .

The aerobatic requirements were completed in June using the O/D "Jigsoar" which is a pod and boom 66 in span A/E/R model with "plug-in" wings and a 30M servo on each aileron. Wing loading is around 13 ounces per square foot. The flights took place in a 20 knot plus easterly – note the plural. The aerobatic schedule, starting with the 2 minutes inverted took about 12 minutes of the specified 1 hour flight. I then set about some lighthearted practice to pass the time. Forty-five minutes into the flight, a slow roll towards the ridge was miscalculated and in the strong wind the model slipped out of the lift and landed out . . . shame.

Casually, I re-launched the glider only to hear Alan Smith, the heartless judge, demand "what are you doing first, the 2 minutes inverted?" Yes, the swine made me start again . . . with friends like that!

In fact the second attempt was in better air and scored marginally more. Having got so far the model was "parked" to save the nicads which had taken a bashing . . . that second hour seemed endless.'.

GLOSSARY OF TERMS

Adverse Yaw The tendency for an aircraft equipped with ailerons to turn in the opposite direction to that intended. It can happen at low speed and is highly likely on high aspect ratio wings.

Aerofoil/Airfoil The shape of the wing when seen in cross-section. A surface that affects the airflow in order to make it produce a useful force, lift, downforce, etc.

Ailerons Movable control surfaces, part of the trailing edge of each wing, used to make the model move in the Roll Axis.

Airbrakes Devices installed in the aircraft structure that will limit the terminal velocity of the model and give extra control over the sink rate. Used for landing and rapid height loss.

Airflow The movement of the air over the aircraft.

Aliphatic Resin A woodworking glue similar to P.V.A., can be sanded more easily. Tends to give a slightly more brittle joint.

All-Moving Tailplane A tailplane where the whole horizontal surface changes angle of attack to act as the elevator.

Altitude Usually taken as the height above the ground, in full size terms usually taken as the height above mean sea level.

Anemometer An instrument for measuring the speed of the wind.

A.M. Type of Radio Control signal, Amplitude Modulation.

Angle of Attack The angle between the Datum Line of the Aerofoil and the Airflow. The angle that the wing meets the air.

Aspect Ratio A ratio found by dividing the mean average chord of the wing into the overall span. Can also be found by dividing the wing area into the square of the span. A high number indicates a long thin wing.

Balance Area The portion of a control surface disposed in front of the pivot point, used to lessen the control force required to move the surface. Generally not desirable on a model.

Ball Connector Type of connector used where the control run ends at a control horn that moves in such a way that the connection has to move in two or more axes, e.g. a steeply raked rudder or a V-tail linkage.

Ballast Weight added to adjust the flying speed and glide angle of the model. See also nose ballast.

Balsa Wood Light South American hardwood used for modelling.

Bank Angle	The angle at which the model is flying to the horizontal plane when it is turning.
B.A.R.C.S.	British Association of Radio Controlled Soarers, the specialist interest body for Soarers. Responsible for most competition classes, achievement schemes, and much more.
Beaufort Scale	A system of describing wind strengths, see Appendix.
Bellcrank	A crank used to change the direction of a control run, usually through 90°
Blue Foam	A polystyrene foam, extruded, it has greater inherent strength than white foam but is heavier. It has a blue coloration that fades upon extended exposure to ultra violet. Trade name "Styrofoam".
Buddy Box	A system of linking two transmitters for training purposes, similar to dual controls on a car.
Bungee	A launching system that uses an elastic member for its motive power. Can be either 'cloth covered" or "surgical".
Camber	The degree of curvature of the wing's mean line, a line drawn equidistant from both surfaces along the length of the chord.
Capstrip	A strip of balsa run along the top or bottom of the rib in the open structure portion of the wing, which adds strength and area for the covering to stick to. Top and bottom capstrips make the rib into an "I" beam.
Carbon Fibre	A man-made fibre used for wing spars and reinforcements in glass-fibre mouldings.
Centre of Gravity (C.G.)	The point at which all the weight of a glider acts, usually taken to mean the point along the root chord of the wing at which the model should balance for correct pitch stability. Denoted on a plan by a black and white crossed circle.
Centre of Pressure (C.P.)	The point on an aerofoil at which all the aerodynamic (lift) forces may be taken as acting. This point will alter with the angle of attack.
Chord	The width of a particular point on the wing, measured from the leading to trailing edge of the wing and parallel to the direction of movement.
Clevis	An adjustable connector used to connect a servo to a control run or a control run to a control surface.
Closed Loop	A control system where two wires are used, one to each side of the control surface and servo arm. The wire in tension actuates the surface, a pull-pull system.
Control Horn	The attachment on a control surface that the control run operates on.
Control Linkage	The attachments on a control system between the control run and the control surface or servo.
Control Run	The system that carries a command from the servo to the control surface.
Convection	The rising of warmed air through cooler air, or any mass through a cooler mass.

Crosswind	A position at right angles to the wind, i.e. to the right or left of the flight line. A line of flight that travels at 90° to the wind.
Crystal	An electrical component used to determine the exact frequency that a transmitter and receiver are working on.
Cumulus	A heaped up cloud with flattish dark base, indicative of thermal activity. May be seen to slowly develop and decay if watched long enough.
Cumulo-Nimbus	A very large cumulus associated with thunderstorms, lift sometimes too strong to be flown in safety.
Cyanoacrylate	"Super-Glue", an adhesive said to have been developed for the sealing of wounds on the battlefield, hence its aptitude for joining fingers. A very useful instant adhesive
Dab	A slang term for a very short application of a control movement, as in "a dab of down".
Datum Line	The line used to construct a drawing around, in our case the important ones are the Aerofoil Section and Fuselage datum lines. Used as a reference for alignment to the airflow, to set the angle of attack.
DEAC	An old fashioned name for rechargeable batteries derived from a tradename, generally refers to flat button style nickel cadmium batteries.
Decoder	Part of the receiver used to translate the control commands.
Definition	The degree of accuracy that a servo control surface can be operated at, measured in +/− degrees.
Dihedral	The angle to the lateral horizontal that the wings are set at, in this case above the horizontal line. If below the line it is known as anhedral. Dihedral imparts a positive lateral stability.
Dope	A cellulose varnish used when covering models in tissue or nylon.
Doubler	A local reinforcement in a structure that increases the thickness of the material and alters its strength, typically used on fuselage sides towards the nose.
Dowel	A round rod usually in wood, often used as a name for a wing joining wire.
Downdraught	A patch of sinking air, colder and denser than the surrounding air.
Downwash	The effect of the wing passing through the air, it deflects the air downwards both in front and behind the wing. Its strength is dependent upon the amount of lift being produced.
Downwind	A position "behind" the pilot's position on the flight line or a course that has the model flying with the wind behind it.
Drag	The resistance of the air to the passage of a body through it. The greater the drag the greater the power required for movement.

Drag Coefficient (cd)	The amount of drag produced by a body as it passes through the air, the lower the better.
Elevator	The control surface that affects the angle of attack of the wing and thus whether the aircraft stalls, flies level or dives, etc. It controls the amount of lift produced and the speed of the model. it works in the pitch axis and may be on the front as in a canard.
Elevons	Found on a flying wing, they perform the function of both the elevator and the ailerons. Special mixing is required.
Epoxy Resin	A two-pack adhesive that cures by an exothermic chemical reaction, can be used for the construction of models or glassfibre mouldings.
F.A.I.	The international body for the control of model flying, mainly concerned with competition rules. Federation Aeronautique International (models section).
Fibreglass	A type of composite material made from glass fibres and a suitable bonding resin. A more accurate name is G.R.P.
Fin	The forward part of the vertical part of the tail from which the rudder is hinged.
Flaps	Movable surfaces on a wing that change the curvature and camber of the wing. They can control the amount of lift and drag produced by a wing section. Commonly used to slow the model when landing, increase the lift on launch and reduce the drag for high speed flight. Positive Flap provides extra lift, Negative Flap reduces drag.
Flaperons	Movable surfaces on a wing that act as both flaps and ailerons.
Flight Line	The position where the pilots stand and fly in a field. At competition it will usually be marked by a line on the ground at right angles to the wind direction.
Flutter	An aerodynamically-produced oscillation of a wing or control surface, similar to the waving and flapping of a flag. Usually happens at high speeds and leads to structural failure.
F.M.	Type of radio control signal, Frequency Modulation.
Former	A structural member used across the fuselage to maintain the shape and rigidity.
Frequency	In general terms refers to the Radio Band used (27, 35 or 459 megaherz). In specific terms refers to the particular spot frequency being used.
Frise Aileron	An aileron that gives an equal drag force in both its up and down movements thus reducing adverse yaw.
F3B	A class of flying governed by the F.A.I., consists of speed distance and duration tasks.
Glass Cloth	A material made from glass fibres woven into cloths with various qualities. Referred to in terms of weight per unit area. Used for glass-fibre moulding and specialised covering.

Glide Ratio	The distance covered by an aircraft for every unit of height lost, i.e. 20:1, 20ft. forward for 1ft. of height lost. It can increase with higher wing loadings for models due to low Reynolds Number effects. The higher the figure the more efficient the model. Often referred to as the "glide angle", a more visual reference.
G.R.P.	Glass Reinforced Plastic, the correct name for glass-fibre.
Heat Shrink Film	The most commonly used form of covering material. Consists of a sheet of plastic with a layer of pigmented heat sensitive adhesive on the reverse side. Is usually ironed on and shrunk taut.
Hi-Start	Another name for a bungee, often used in European and American kit instructions.
Incidence	Usually talked of as the "angle of incidence". The angle at which the datum line of the wing section (aerofoil) is set in relation to the datum line of the fuselage. It is common practice to use the bottom surface of the wing as a datum line for this purpose, and while this may work it is better to use the wing performance data and set according to the needs of the model using the datum lines.
Included Angle	The angle between the halves of a V-tail, usually ranging from 90 to 110°.
Kevlar	A synthetic Aramid fibre of great strength used in G.R.P. mouldings, blunts cutting tools very quickly and does not sand, preferring to fray.
Kinetic Energy	The energy that a body possesses due to its motion and mass.
Knuckle Hinge	A hinge that has a rounded front profile and turns in a semicircular cut-out, pivoted at the centre of the semi-circle. Minimises drag and the air gap.
Laminar Flow	A smooth uninterrupted airflow that is not turbulent.
Landing Circle	A circle of 25m diameter used in Open Class competitions to determine additional points and ensure tidy operation of the flying. 50 seconds are added to the flight time for landing in the "circle".
Lateral	A direction across the span of the model, at right angles to the fuselage centre line. Hence Lateral Control is achieved in the roll axis by the use of ailerons.
League Contest	An open contest the results of which count towards the local leagues for Open competitions run by BARCS.
Leading Edge (l.e)	The front edge of a flying surface.
Lenticular Clouds	Clouds formed downwind of large hills and mountains, aligned across the wind. They are formed by the leewaves.
Lift	The force created by a flying surface as it moves through the air at an angle of attack, is generally at right angles to the

surface and can be up or downwards in direction. The flying attitude of the model is not relevant to the lift, it can be produced in a vertical climb or dive as easily as in straight and level flight.

Lift/Drag Coefficient (cl/cd)
A term denoting the relationship between the lift and the amount of energy required to produce it at a particular angle of attack. The lower the figure the better.

Longeron
A structural member running along the length of the fuselage.

Longitudinal
A direction along the length of the fuselage, down the centre line.

Maximum
The score given to the winner in a percentage slot or the completion of the complete time in a time task flight.

Mean Average Chord (m.a.c.)
The average chord of the wing. On straight taper wings found by dividing the sum of the root and tip chords by 2. On a complex tapered wing found by dividing the wing area by the span.

Minimum Sink (min. sink)
The trim for the maximum flight time from any given height. The lowest rate of sink possible with the model.

Mixer
A mechanical or electrical device that allows two functions to be performed by a control surface, elevons, flaperons, etc.

Monofilament Line
Nylon fishing line, used for model towing.

Multi Task
Another name for F3B style competitions.

Mylar
A plastic material in strip form that is used for hinges and control surface gap sealing.

Nationals
The championships run by the SMAE.

Nicad
Abbreviation from Nickel Cadmium. A rechargeable battery used in models.

Nose Ballast
The weight put in the nose of the aircraft to balance the model at its recommended centre of gravity position.

Nose Moment
The distance from the tip of the nose to the centre of gravity. A long nose moment reduces the amount of nose ballast required.

Nylon
A material in cloth form used to cover models when high strength is required.

Percentage Slot
A form of scoring where the winner is given 1000 points and the rest of the fliers in the time slot are marked as a percentage of the winner's score.

Piano Wire
A high carbon steel wire with springy qualities used for wing joiners.

Pitch
In airframe terms, the nose-up or nose-down movement of the model, controlled by the elevator. Movement about the lateral axis.

Planform	The shape of the wing (or whole model) when viewed from directly above.
Pod and Boom	Type of fuselage using a thinner tube aft of the wing trailing edge. Often supplied as a separate boom and front end pod.
Polyester Resin	Commonly used resin for glass-fibre work. Two part resin, it cures by a chemical reaction liberating heat, an exothermic reaction; beware of over-enthusiastic use of the catalyst.
Positive Camber	A curved mean line where the curvature is above the wing's datum line.
Postal Competition	A competition open to all comers, run during a set period at any field convenient to the competitor, the results being sent to the organisers by post.
Pot	Common name for a Potentiometer, a variable resistance device used to control the position of a servo arm. It can be subject to wear after some time leading to a jumpy servo. Also a trophy won at a competition.
Potential Energy	The energy that a model possesses as a result of its position (height).
Precision Task	A flight of prescribed duration tied to a landing task, used in F3B.
Projected Area	The area of a wing when viewed from directly above, no allowance being made for dihedral, etc. Can be applied to V-tails.
Proportional Control	The system used on all modern radio control sets, where the servo moves to a position dictated by the position of the control stick.
Pushrod	A control linkage consisting of a rigid rod, forms a push-pull system.
PVA Glue	Poly Vinyl Acetate, a flexible glue used for wooden parts of the airframe. Used for light to medium stress joints.
Radioglide	Three day competition run by BARCS, the major competition soaring event of the year.
Releasable Towhook	A type of towhook most suited to operation from the bungee. Requires careful setting up, or it may not release the line upon the command.
Rib	The structural component that lies across the width of the wing and gives it its characteristic shape.
Roll	A lateral rotation about the fuselage centre line, the longitudinal axis. As an aerobatic manoeuvre it involves a full 360 degree movement.
Rudder	The vertical control surface attached to the fin. it controls the yaw axis of the model, assisting the turn on an aileron model and controlling it on a rudder/elevator model.
Ruddervators	The combined rudders and elevators found on a V-tail.
Rx	A common abbreviation for receiver.

Sailplane	The name given to a glider of high performance.
Separated Flow	See turbulent flow.
Servo	The electro-mechanical device in the model that controls a control surface.
Shoulder Mounted Wing	A wing mounted on top of, or level with the top of, the fuselage.
Sink	Air that is falling towards the ground due to its greater density, normally colder than the surrounding air, to be avoided at all costs.
Sink Rate	The height that the model drops for every second of its flight. Measured in ft/sec. or metres/sec. Affected by atmospheric conditions and trim.
Sliding Servo	A system often used on Flaperons, Elevons and Rudder-vators, where one servo is mounted on a sliding tray, the movement of which is controlled by another servo.
Slop	Free movement in a control system, felt as play on the control surface. This reduces the response speed of the model and makes trimming difficult. Avoid at all costs.
Slope Soaring	The flying of models in air that is rising as it travels over geographical features such as hills ridges, etc.
S.M.A.E.	The Society of Model Aeronautical Engineers, the officially recognised official body for the control of model flying in general in Britain.
Snake	Common name for a push-pull cable linkage to a control surface. Several types available.
Span	The dimension of a wing or tailplane when measured from tip to tip, or any other such surface when measured at right angles to the airflow.
Spar	A structural component in a flying surface that is situated along the span and carries the major flight loads.
Spoilers	Originally a device for disrupting the airflow over a wing and thus allowing control of the sink rate by reducing lift. Now commonly accepted as the name for the "letter box" style airbrakes fitted to many models, which will also act as drag brakes and limit the speed of the model. Thus they operate as both spoilers and airbrakes.
Spruce	Light strong wood used for parts of the airframe that carry medium to high stress. Technically a softwood.
Stall	Technically the separation of the airflow from the airfoil, can occur on any surface. In model terms the manoeuvre that results from trying to fly too slowly.
Standard Class	Another name for 100S competition model, restricted to 800 sq. in. wing area and 100 in. span, only rudder airbrake and elevator allowed. In America the wing area is set at 900 sq. in.
Stick	Common name for a control stick, derived from the full size term "joystick".
Stressed Skin	A type of structure where the stresses are carried in the

Micro radio is used in this hand-launched glider, *Double Trouble* **by Eddie Walker**, which incorporates elliptical dihedral. These models can pick up rising air from only a few feet altitude.

external skin of the component, e.g. foam wings, GRP mouldings.

Styrofoam
"Blue Foam", the trade name for extruded polystyrene foam, has better structural qualities, particularly in compression.

Symmetrical Section
An aerofoil that has a flat camber line and a streamline form of equal thickness either side.

Tail Moment
The distance between the wing and the tailplane. Often taken as being the distance between the trailing edge of the wing and the leading edge of the tailplane. This is only a guide, the moment arm should be taken from the 25% M.A.C. point on the wing to the 25% M.A.C. point on the tailplane.

Tailplane
The flying surface that controls the model in pitch, i.e. diving or climbing. More accurately referred to in America as a horizontal stabiliser or "stab".

Taper
A measure of the degree of narrowing of a flying surface, taken from the root to the tip. Found by dividing the tip chord by the root chord, can be expressed as a single i.e. 0.6 or a ratio 0.6:1. For general sport flying a strong taper is to be avoided as it can lead to tip stalls.

Terminal Velocity
The maximum speed that the model is capable of travelling at in a dive. Seldom reached, structural failure occurs or the ground intervenes first with normal models. It can be lessened by the use of airbrakes.

Thermal
A mass of air warmer than the surrounding air that rises due to its ligher weight, a similar effect to that used by hot air balloons.

Thermal Turn
A turn hopefully performed in a thermal, the model being banked over and turned in continuous circles while trimmed for minimum sink.

Tissue
A paper type material used in conjunction with Cellulose Dope as a covering material.

Torsional Stiffness
The ability of a structure to resist twisting, of particular importance on wings and tailplanes.

Towline
Usual name for the apparatus used to hand tow models, also the line itself.

Trailing Edge
The rear edge of a flying surface. Often written as t.e. for short.

Trim
The setting of the model in a particular state where the aerodynamic forces are in equilibrium, e.g. minimum sink, shallow dive, etc. Also the common name for the control at the side of the control stick that allows one to apply fine adjustments to the centring of the servos and thus change the flying trim of the model.

Trimming
The process of setting the model up to fly in a satisfactory fashion according to the pilot's preferences.

Two Metre Class
A restricted class for rudder/elevator models with a maximum wing span of 2m. In America no limit is applied to the control functions.

Tuck Under	An unpleasant condition where the model enters an ever steepening dive that is not recoverable with up elevator, try down, gently. Often caused by a wing with too little torsional stiffness.
Turbulent Flow	A form of airflow characterised by the mixing of the layers in the boundary layer close to the wing. Compare with laminar flow.
Tx	A convenient abbreviation for the word Transmitter.
U.H.F.	A frequency for radio control purposes, Ultra High Frequency. 459 megaherz being used in Britain.
Upwind	A position that is in front of the flight line, the direction from which the wind is coming, and a course that is directly into the wind.
Variable Camber	The use of flaps to alter the camber line of the aerofoil and thus its flight characteristics.
V.N.E.	Velocity Never to be Exceeded, a full size term giving a value to the maximum safe speed for the aircraft, a useful concept for models but hard to apply.
Veneer	A thin sheet of wood used to make the external surface of a wing, usually used to refer to the Obechi veneer used on foam wings. The veneer can be balsa or any other suitable wood.
Vortices	The circular twisting airflows found in the wake of a wing or flying surface. Their intensity is dependent upon the amount of lift being produced.
Warp	An unwanted twist in a structure.
Wash In	A twist in the wing such that the tip of the wing has a greater angle of attack than the root. Highly undesirable on a normal style of wing.
Wash Out	A twist in the wing such that the tip of the wing has a lower angle of attack than the root. Can be used as a means of preventing tip stalls, but is less efficient overall as it promotes high drag at high speeds and can lead to "tuck under" problems.
Wave	An atmospheric phenomenon found in the lee of hills, much like the waves seen behind a stone in a river. Flying in the rising part of the wave will yield continuous lift conditions. Wave lift is usually only contacted at a fair height and is often marked by lenticular clouds.
Web	A structural component used to connect the top and bottom spars, it controls the transmission of compression loads from the top spar.
Wind Shadow	A geographical feature such as a hollow or area shielded by a bank of trees. Air in such areas can warm up and form potential thermals.
Wind Shear (Gradient)	The tendency of the air to travel slower the nearer it is to the ground. Of importance when landing and more pronounced in high winds.

Wing Area	The surface area of one side of the wing, found by multiplying the span by the mean average chord. Units: square feet, square inches or square decimetres.
Wing Loading	The amount of weight that each unit area of the wing must carry to sustain flight. Found by dividing the wing area into the overall weight. Units: ounces/square foot or grammes/square decimetre.
Wing Section	The cross sectional shape of the wing, interchangeable with the terms wing profile, aerofoil section or airfoil section.
Wing Thickness	The depth of the wing at its maximum thickness expressed as a percentage of the wing chord, usually around 10%.
Winch	A device for launching models with a line, the Hand Winch being run with by the tower and the Power Winch being fixed to the ground and run by electricity.
Working Time	The time allowed for the completion of a task in competition.
Xtal	Common abbreviation for a radio crystal.
Yaw	A movement caused by use of the rudder, the fuselage rotating about its centre of gravity like a weathercock.
Yaw/roll	The result of holding the rudder on with a dihedral wing model. The model starts to roll and changes direction.

INDEX

Ballasting 135–136
Ballast stowage 48–49
Batteries 14–15
Bungees 16–17, 67–69
Bungee techniques 59–60, 62

Centre of gravity 59–60, 62
Clothing 18–19
Clubs 9
Cold weather 143
Colour schemes 49–50
Competitions (slope) 163–166
   Club events 166–168
   Cross country 166–167
   National events 166–168
   Multi-task 167
   Precision aerobatics 167
   Pylon racing 167
   Scale 168
Competitions (thermal) 144–160
   Alternatives 160
   Case for 144–147
   First steps 147–148
   Fly off 159–160
   Open 151–154
   Tactics & techniques 154–157
   Team functions 148–151
Control movements 60
Control runs 27–34
   Cable linkages 28–29
   Closed loop linkages 30–33
   Pushrod linkages 29–30
Control throw adjustments 63–64
Covering 45–48
   Glassfibre 47–48
   Heat shrink 46–47
Crash inspection 56–58
Crash padding 25–27

Dive recovery 84
Double centring 33–34

"Essing" 93, 138

Fin & rudder 39

Finishing, Glassfibre fuselages 48
Flutter 44–45
Foul weather flying 133–138
Fuselages 21–25
   Glassfibre 23–25
   Wooden 22–23

Groundspeed effects 136–138

Handtowing 73–82
High speed flying 88
Hinges 40–42
Hi-Start 16–17

Inconsistent centring 33–34
Insurance 19

Landing (slope) 123, 128–131
   Circular approach 129
   Figure of eight approach 123
   Slope side approach 130–131
   Square approach 130
Landing (flat field) 91–97
   Airbrakes 94
   Alternatives 96
   Approach 91–92
   Downwind leg 93
   Landing window 94–95
   Obstacle avoidance 97
   Overshooting 93–94
   Touching down 95
   Turns 96
   Undershooting 93
Launching (flat field) 67–82
Launching (slope) 124
Lift recognition 100–105
Loops 89
Low cloud 142

Nicads 14–15

Open class 151–154
Orientation 87–88
Overhead flying 88–89

Power winches 80–82

Radio sets 11–16
Rain 141–142
Repairs 51–58
    Built Up wings 55–56
    Foam wings 53–55
    Fuselages 51–53
    Tailplanes 56
Rolls 89

Servo installation 25
Servo output adjustment 64–65
Slope sites 111–114
Slope soaring 108–131
    Advanced techniques 131–132
    Conditions 114–116
    Equipment 108–111
    Flying skills 124–128
    Initial survival 120–122
    Learning solo 116–117
    Model preparation 117–120
Spins 88
Spiral dive 88
Stall recovery 83–84
Static balancing 59–60
Straight & level flying 84–86

Tailplane 36–38
Tailplane cranks 38
Test gliding 61–62
Thermalling 98–107
    Descent planning 106–107
    Equipment 98
    Maximising lift 105–106
    Preliminary work 99
    Thermal signs 100–105
Thermal turns 87
Tip stall recovery 88
Tools 17–18
Towhooks 34–36
    Fixed 35–36
    Releasable 34–35
Towhook adjustments 65–66
Towline launching 73–82
    Pilots techniques 74–75
    Towmans techniques 76–79
    Windspeed effects 78
Trainers 10–11, 108–111
Transmitter adjustments
    (trimming) 65
Trimming 59–66
"Tuck under" 44–45, 134
Turbulation 66
Turbulence 139–140
Turning 86–87

Visibility 142–143

Wind shear 140–141
Windspeed effects on
    bungees 72–73
Windy weather flying 133–141
Wings 42–45
    Built up 42–44
    Foam 42
    Setting jig 45
    Torsional stiffness 44–45, 134
Wing sections (modern) 135

Zooming 87

OTHER MODELLING TITLES
IN THE
ARGUS RANGE INCLUDE:

Four-stroke Handbook

Basic Aeromodelling

Fifty Years of 'Aeromodeller'

Radio Control Primer

The Buggy Book

The Buggy Racing Handbook

Boat Modelling

Handbook of Ship Modelling

Model Yachting

Radio Control in Model Boats

Historic Ship Models

Introducing R/C Model Boats

Military Modelling Guide to Military Modelling

Painting and Finishing Models

Model Flight

Model Aeroplane Building Sketch by Sketch

Introducing Model Aero Engines

Model Aircraft Aerodynamics

Introducing R/C Model Aircraft

Scale Model Aircraft for Radio Control

Building & Flying R/C Model Aircraft

Manual of Electric R/C Cars

R/C Model Racing Cars

Introducing R/C Model Boats

R/C Fast Electric Power Boats

Introducing Model Marine Steam

Airbrushing & Spray Painting Manual

Radio Controlled Gliding

Military Modelling Guide to Wargaming

Military Modelling Guide to Solo Wargaming

Introduction to Electric Flight

plus many model engineering and workshop practice books

Argus Books Limited
Wolsey House, Wolsey Road, Hemel Hempstead
Hertfordshire HP2 4SS

Send S.A.E. for latest book list

3 GOOD REASONS..

..for subscribing!

Every month, these three informative magazines provide the expertise, advice and inspiration to keep you abreast of developments in the exciting field of model aviation.

With regular new designs to build, practical features that take the mysteries out of construction, reports and detailed descriptions of the techniques and ideas of pioneering aircraft modellers all over the world, they represent three of the very best reasons for taking out a subscription. So that you need never miss a single issue or a single minute of aeromodelling pleasure!

	UK	EUROPE	MIDDLE EAST	FAR EAST	REST OF THE WORLD
Radio Modeller	£16.80	£22.40	£22.60	£24.80	£23.00
RCM&E	£16.80	£22.80	£23.00	£25.80	£23.40

Airmail rates given on request

Your remittance with delivery details should be sent to:
The Subscription Manager (CG49)
Argus Specialist Publications
Argus House
Boundary Way
Hemel Hempstead
Herts HP2 7ST